# CAMBRIDGE TRACTS IN MATHEMATICS

GENERAL EDITORS

H. BASS, J. F. C. KINGMAN, F. SMITHIES,
J. A. TODD & C. T. C. WALL

---

## 66.  *Fixed point theorems*

## D. R. SMART

*Senior Lecturer in Mathematics*
*University of Cape Town*

# *Fixed point theorems*

**CAMBRIDGE UNIVERSITY PRESS**

Published by the Syndics of the Cambridge University Press
Bentley House, 200 Euston Road, London NW1 2DB
American Branch: 32 East 57th Street, New York, N.Y.10022

© Cambridge University Press 1974

Library of Congress Catalogue Card Number: 73–79314

I S B N: 0 521 20289 2

First published 1974

Printed in Great Britain
at the University Printing House, Cambridge
(Brooke Crutchley, University Printer)

# Contents

[ v ]

# Preface

This book is intended as an introduction to fixed point theorems and to their applications in analysis. Apart from applicable theorems, I have included those which interested me.

Since applications usually involve spaces of functions, I give Banach space versions of most of the theorems. The book is thus aimed at readers with a general interest in functional analysis. However, I have hardly touched on a series of recent developments, by F. E. Browder and others; for these see Browder's forthcoming book. To fill obvious gaps in the direction of pure topology, the reader can refer to the excellent surveys by van der Walt (1963), Bing (1969), and Fadell (1970), and to Brown's book (1970).

The methods of proof are those which will seem natural to the functional analyst. Most of the results are derived from Brouwer's fixed point theorem for the ball $B^n$; to get this theorem we use some facts about homology groups. After that, algebraic topology is rarely mentioned. I have chosen to give geometric proofs, rather than that to develop degree theory and base everything on that. The degree and other invariants are discussed at the end of the book.

I must thank Dr F. Smithies for suggesting that I write this book, and for helpful comments on successive versions of it. I am also indebted to the University of Cape Town for two periods of study leave during which much of the writing was done, and to the Universities of Cambridge, Edinburgh and Wales (Swansea) for their hospitality on these occasions. I should mention, too, the contribution of various members of those universities who attended my lectures at a time when I was still sorting out my ideas.

However, my greatest helper has been Daphne, who provided conditions in which I felt like working, and who spent many hours typing the manuscript.                      D. R. SMART

*Cape Town*
*March* 1973            [ vii ]

# Symbols used

| | |
|---|---|
| $\mathcal{M}^0$ | Interior of a set $\mathcal{M}$ |
| $\overline{\mathcal{M}}$ | Closure of $\mathcal{M}$ |
| $\partial\mathcal{M}$ | Boundary of $\mathcal{M}$ |
| $\mathrm{co}(\mathcal{M})$ | Convex cover of $\mathcal{M}$ |
| $\overline{\mathrm{co}}\,(\mathcal{M})$ | Closed convex cover of $\mathcal{M}$ |
| $C\,(\mathcal{M})$ | Space of continuous bounded functions on $\mathcal{M}$ |
| $\chi(\mathcal{M})$ | Measure of noncompactness (p. 32) |
| $\mathrm{span}\,(\mathcal{M})$ | Linear subspace spanned by $\mathcal{M}$ |
| $l^2$ | Space of square-summable sequences |
| $\mathcal{H}_0$ | Hilbert cube (p. 13) |
| $\mathcal{D}(T)$ | Domain of a mapping $T$ |
| $\mathcal{R}(T)$ | Range of $T$ |
| $\mathcal{G}(T)$ | Graph of $T$ |
| $F(T)$ | Set of fixed points of $T$ |
| $N(x,\epsilon)$ | $\epsilon$-neighbourhood of $x$ |
| $(\cdot\,,\cdot\,)$ | Inner product |
| $\varnothing$ | Empty set |
| $I$ | Identity operator |
| $\mathcal{B}^*$ | Dual of Banach space $\mathcal{B}$ |
| $Z$ | Group of integers |
| $\deg(\ \ )$ | Degree (pp. 77, 80) |
| $\mathrm{rot}(\ \ )$ | Rotation (p. 75) |

# 1. *Contraction mappings*

## 1.1  Introduction

Consider a mapping $T$ of a set $\mathcal{M}$ into $\mathcal{M}$ (or into some set containing $\mathcal{M}$). One of the few questions we can ask (in this general setting) is whether some point is mapped onto itself; that is, does the equation

$$Tx = x$$

have a solution? If so, $x$ is called a *fixed point* of $T$. The theorems we prove assert that, under suitable conditions on $\mathcal{M}$ and $T$, a fixed point exists.

Obviously, the conditions must always imply that $\mathcal{M} \neq \varnothing$. Usually, $\mathcal{M}$ is a topological space and some conditions of continuity and compactness (or at least completeness) are needed.

We shall see that many existence theorems of analysis can be treated as special cases of suitable fixed point theorems.

In the present chapter, we place rather strong conditions on $T$ and rather weak conditions on $\mathcal{M}$. Because of the simplicity of its assumptions, 1.2.2 is the most widely applied fixed point theorem. We discuss some of its applications in §§1.3, 1.4, 6.2 and 6.5.

We first give a few simple and general results.

THEOREM 1.1.1  *If $T$ maps $\mathcal{M}$ into $\mathcal{M}$ then any fixed point $z$ of $T$ is in $\cap\, T^n\mathcal{M}$. Conversely, if $\cap\, T^n\mathcal{M} = \{y\}$, a one-point set, then $y$ is a fixed point for $T$.*

*Proof.* Since $Ty$ must be in $\cap\, T^n\mathcal{M}$, we have $Ty = y$. $\square$

THEOREM 1.1.2  (*Principle of successive approximations*) *If $T$ is continuous on a Hausdorff topological space $\mathcal{M}$ to $\mathcal{M}$ and if $\lim T^n x = y$ exists then $Ty = y$.*

*Proof.* $Ty = T(\lim T^n x) = \lim T^{n+1} x = y$. $\square$

[ 1 ]

THEOREM 1.1.3  *Let $\mathscr{U}$ be a metric space. Suppose that $T$ is a continuous mapping of (a closed subset of) $\mathscr{U}$ into a compact subset of $\mathscr{U}$ and that, for each $\epsilon > 0$, there exists $x(\epsilon)$ such that*

$$\rho(Tx(\epsilon), \quad x(\epsilon)) < \epsilon. \tag{1}$$

*Then $T$ has a fixed point.*

*Proof.* Let $T$ map the closed subset $\mathscr{M}$ into the compact subset $\mathscr{L}$. Since $Tx(\epsilon)$ is in $\mathscr{L}$ we can assume that for some sequence $\epsilon_n \to 0$ we have $Tx(\epsilon_n) \to y \in \mathscr{L}$. By (1) we also have $x(\epsilon_n) \to y$ so that $y \in \mathscr{M}$. Thus $Ty$ is defined and

$$Ty = T(\lim x(\epsilon_n)) = \lim Tx(\epsilon_n) = y. \quad \square$$

DEFINITION 1.1.4  The points $x(\epsilon)$ satisfying (1) will be called *$\epsilon$-fixed points* for $T$.

We shall often use 1.1.3 to obtain fixed points from $\epsilon$-fixed points. However, the usual position is that we can obtain the $\epsilon$-fixed points by a constructive argument. Theorem 1.1.3 does not give a constructive proof for the existence of fixed points. Brouwer (1952) argues that only $\epsilon$-fixed points have meaning for the intuitionist.

## 1.2   The contraction mapping theorem

DEFINITION 1.2.1  Let $T$ be a mapping of a metric space $\mathscr{M}$ into $\mathscr{M}$. We say that $T$ is a *contraction mapping* if there exists a number $k$ such that $0 < k < 1$ and

$$\rho(Tx, Ty) \leqslant k\rho(x, y) \quad (\forall x, y \in \mathscr{M}). \tag{1}$$

The following result is called the *Contraction Mapping Theorem*.

THEOREM 1.2.2 (Banach, 1922)  *Any contraction mapping of a complete non-empty metric space $\mathscr{M}$ into $\mathscr{M}$ has a unique fixed point in $\mathscr{M}$.*

*Proof.* Let the mapping $T$ satisfy (1) for some $k < 1$. Choose any point $y$ in $\mathscr{M}$. The sequence of points $T^n y$ satisfies, for $n > 0$,
$$\rho(T^n y, T^{n+1} y) \leqslant k\rho(T^{n-1} y, T^n y),$$
so that by induction

$$\rho(T^n y, T^{n+1} y) \leqslant k^n \rho(y, Ty).$$

By the triangle inequality we have for $m \geqslant n$

$$\rho(T^n y, T^m y) \leqslant \rho(T^n y, T^{n+1} y) + \rho(T^{n+1}, T^{n+2} y) + \ldots + \rho(T^{m-1} y, T^m y)$$
$$\leqslant (k^n + k^{n+1} + \ldots + k^{m-1}) \rho(y, Ty)$$
$$\leqslant k^n (1-k)^{-1} \rho(y, Ty). \tag{A}$$

Thus $\rho(T^n y, T^m y) \to 0$ if $m, n \to \infty$. Since $\mathcal{M}$ is complete the sequence $T^n y$ has a limit $z$ in $\mathcal{M}$. By 1.1.2, $z$ is a fixed point for $T$. This fixed point is unique since if $Tz = z$ and $Tw = w$ we have

$$\rho(z, w) = \rho(Tz, Tw) \leqslant k\rho(z, w)$$

so that $\rho(z, w) = 0$; that is, $z = w$. $\square$

For applications of 1.2.2, some further facts are important.

REMARK 1.2.3 *Under the conditions of* 1.2.2:

(i) *the fixed point $z$ can be calculated as* $\lim T^n y$ *for any $y$ in $\mathcal{M}$*;

(ii) $\rho(T^n y, z) \leqslant k^n (1-k)^{-1} \rho(y, Ty)$;

(iii) *For any $y$ in $\mathcal{M}$*, $\rho(y, z) \leqslant (1-k)^{-1} \rho(Ty, y)$.

*Proof.*

(i) is clear from the proof of 1.2.2;

(ii) follows by letting $m \to \infty$ in the inequality (A);

(iii) follows from (ii) or from the inequality

$$\rho(y, z) \leqslant \rho(y, Ty) + \rho(Ty, Tz) \leqslant \rho(y, Ty) + k\rho(y, z). \quad \square$$

There is an alternative form of 1.2.2 in which the contraction mapping is only defined on a suitable neighbourhood of the point $y$ which is taken as the first approximation. This is suggested by 1.2.3 (iii), which gives a neighbourhood of $y$ in which the fixed point must lie. For details and applications of this alternative theorem, see Copson (1968).

## 1.3 The Cauchy–Lipschitz theorem

We use the contraction mapping theorem to establish an existence–uniqueness theorem for ordinary non-linear differential equations.

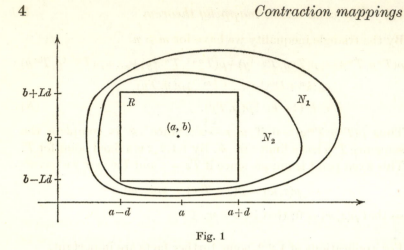

Fig. 1

**THEOREM 1.3.1** (Lipschitz, 1876)  *Let $f$ be continuous and satisfy a Lipschitz condition with respect to $y$:*

$$|f(t,y)-f(t,z)| \leqslant K|y-z|$$

*in some neighbourhood $N_1$ of $(a,b)$. Then the differential equation with initial condition*

$$\frac{dy}{dt} = f(t,y), \quad f(a) = b \tag{1}$$

*has a unique solution in some neighbourhood of $a$.*

*Proof.*  We observe that (1) is equivalent to the integral equation

$$y(t) = b + \int_a^t f(x, y(x))\,dx \tag{2}$$

(for an approach which makes this transformation of the problem seem less accidental, see Chapter 5). We consider a set $\mathcal{M}$ of functions, and a mapping $U$ in $\mathcal{M}$. The image $Uy$ of a function $y$ with values $y(x)$ will be given by

$$(Uy)(t) = b + \int_a^t f(x, y(x))\,dx. \tag{3}$$

How can we find a set of functions which is mapped into itself by $U$? We first choose a compact neighbourhood $N_2$ of $(a,b)$, inside $N_1$; then $f$ is bounded on $N_2$, say

$$|f(x,y)| \leqslant L \quad ((x,y) \in N_2).$$

If $y$ is a function with graph in $N_2$ we have

$$|Uy(t) - b| = \left| \int_a^t f(t, y(t))\, dt \right| \leqslant L|t - a|.$$

This means that if $y$ is a continuous function defined for

$$|t - a| \leqslant d,$$

for which $|y(t) - b| \leqslant Ld$, then $Uy$ satisfies the same conditions. We must choose $d$ small enough for the rectangle (figure 1)

$$R = \overline{N}(a, d) \times \overline{N}(b, Ld)$$

to be in $N_2$. We then define $\mathcal{M}$ to be the set of continuous functions with graphs in $R$, and our argument shows that $\mathcal{M}$ is mapped into itself by $U$. We use the upper bound norm on $\mathcal{M}$.

To ensure that $U$ is a contraction mapping we should also arrange, in choosing $d$, that $dK < 1$. Then we have, for $y$ and $z$ in $\mathcal{M}$

$$\begin{aligned}
|Uy(t) - Uz(t)| &= \left| \int_a^t f(x, y(x)) - f(x, z(x))\, dx \right| \\
&\leqslant d \sup |f(x, y(x)) - f(x, z(x))| \\
&\leqslant dK \sup |y(x) - z(x)|.
\end{aligned}$$

Thus

$$\begin{aligned}
\|Uy - Uz\| &= \sup_t |Uy(t) - Uz(t)| \\
&\leqslant dK \sup |y(x) - z(x)| = dK \|y - z\|,
\end{aligned}$$

and since $dK < 1$, $U$ is a contraction mapping. Thus by 1.2.2, $U$ has a unique fixed point in $\mathcal{M}$. This means that there is a unique function in $\mathcal{M}$ which is a solution of (1). Since any solution of (1) is in $\mathcal{M}$ (for $d$ sufficiently small), there is a unique solution of (1).  $\square$

## 1.4   Implicit functions

We give a second application of the contraction mapping theorem.

THEOREM 1.4.1 (*Implicit Function Theorem*)   *Let $N$ be a neighbourhood of a point $(a, b)$ in $R^2$. Suppose that $f$ is a continuous* *function of $x$ and $y$ in $N$ and that $\partial f/\partial y$ exists in $N$ and is continuous at $(a, b)$. Then if*

(i) $\dfrac{\partial f}{\partial y}(a, b) \neq 0$,

(ii) $f(a, b) = 0$,

*there is a unique continuous function $y_0$ on some neighbourhood of $a$, such that $f(x, y_0(x)) = 0$.*

*Proof.* We write $D_f$ for $\partial f(a, b)/\partial y$. We will look for a fixed point of a mapping defined by

$$Tz(x) = z(x) - D_f^{-1} f(x, z(x)).$$

(This mapping is suggested by the idea of finding $y_0(x)$ by Newton's method.) It is clear that if $y$ is fixed we must have $f(x, y(x)) \equiv 0$. We will find a set of functions $\mathcal{M}$ such that $T$ maps $\mathcal{M}$ into $\mathcal{M}$ and that $T$ is a contraction mapping in $\mathcal{M}$. Within $N$ we choose a closed rectangle

$$R = \overline{N}(a, \epsilon) \times \overline{N}(b, \delta)$$

small enough to give

$$\left| D_f^{-1} \frac{\partial f}{\partial y}(x, y) - 1 \right| < \tfrac{1}{2} \quad ((x, y) \in R),$$

$$\left| D_f^{-1} f(x, b) \right| < \tfrac{1}{2}\delta \quad (|x| \leqslant \epsilon).$$

Now write $C = C(\overline{N}(a, \epsilon))$ and put

$$\mathcal{M} = \{ y \in C : y(a) = b, \quad \| y - \beta \| \leqslant \delta \}$$

(where $\beta$ is the function identically equal to $b$). Clearly $T$ maps $\mathcal{M}$ into $C$. We have

$$\| T\beta - \beta \| = \| D_f^{-1} f(x, b) \| < \tfrac{1}{2}\delta.$$

For $(x, y)$ in $R$ we have

$$\left| \frac{\partial}{\partial y}(y - D_f^{-1} f(x, y)) \right| = \left| \left( 1 - D_f^{-1} \frac{\partial}{\partial y} f(x, y) \right) \right| < \tfrac{1}{2}.$$

Thus by the lemma below, if $y$ and $z$ are in $\mathcal{M}$,

$$|Ty(x) - Tz(x)| \leqslant \tfrac{1}{2} |y(x) - z(x)| \quad (x \in \overline{N}(a, \epsilon)),$$

so that $\|Ty - Tz\| \leqslant \frac{1}{2}\|y - z\|$. Thus $T$ is a contraction mapping. Also

$$\|Ty - \beta\| \leqslant \|Ty - T\beta\| + \|T\beta - \beta\|$$
$$\leqslant \tfrac{1}{2}\|y - \beta\| + \|T\beta - \beta\|$$
$$< \tfrac{1}{2}\delta + \tfrac{1}{2}\delta = \delta$$

so that $T$ maps $\mathscr{M}$ into $\mathscr{M}$. Since $\mathscr{M}$ is complete, $T$ has a unique fixed point in $\mathscr{M}$. Thus our problem has a unique solution which can be calculated by successive approximations, using the operator $T$ and starting from any member of $\mathscr{M}$. □

LEMMA *If* $|\partial f/\partial y| \leqslant \frac{1}{2}$ *at all points between* $(x, y)$ *and* $(x, z)$ *then* $|f(x, y) - f(x, z)| \leqslant \frac{1}{2}|y - z|$.

*Proof.* Use the mean value theorem. □

The argument given above can be used to prove a far more general form of the implicit function theorem. If $f$ maps $\mathscr{B} \times \mathscr{C}$ into $\mathscr{C}$, where $\mathscr{B}$ and $\mathscr{C}$ are Banach spaces we must interpret $D_f$ as the Fréchet derivative of $f$ at $(a, b)$; we replace (i) by the condition that $(D_f)^{-1}$ exists. We interpret $\mathscr{M}$ as a space of continuous functions defined on a neighbourhood in $\mathscr{B}$ with values in $\mathscr{C}$. Since the lemma remains true for the Fréchet derivative, all details of the proof carry over directly to the general case. In particular, if $f$ maps $R^m \times R^n$ into $R^n$ we interpret $D_f$ as an $n \times n$ matrix of partial derivatives with respect to the $y$-variables and replace (i) by the condition that this matrix has an inverse.

In the above argument each value of $y_0(x)$ is calculated by a variant of Newton's method. It is easy to adapt the argument to show that for each $x$ in a neighbourhood of $a$, the value $y_0(x)$ (such that $f(x, y_0(x)) = 0$) can be found by the ordinary Newton's method. Thus the conditions of the theorem give sufficient conditions for Newton's method.

## 1.5   Other applications of Banach's theorem

Various applications of the contraction mapping theorem are given in Kolmogorov and Fomin (1957). These provide excellent illustrations of the use of fixed point theorems in analysis.

It is sometimes doubtful whether to refer to the contraction mapping theorem in a proof, or to carry out the discussion in terms of successive approximations. In linear problems we could just as well use the Neumann series (for example, in the applications in Kolmogorov and Fomin; or in Harris, Sibuya and Weinberg (1969)). On the other hand, in the discussion of the Schwarz alternating method in Courant and Hilbert (1962, p. 293), it is clear that the discussion could have been phrased in terms of a contraction mapping but in fact the convergence of the successive approximations is discussed directly. The only disadvantage – virtual repetition of the proof of the contraction mapping theorem – is slight (since this proof is short) and is compensated by the advantage of having an argument complete in itself.

In the case of the more difficult fixed point theorems which we will give later, there is a definite gain when the theorem is applied, since by appealing to a general theorem which depends on a deep argument, we can hope to avoid going through an argument of similar depth in the particular case to which the theorem is applied.

## Exercises

1. Show (by 1.2.2) that there is a unique continuous function $f$ on $[-1, 1]$ such that

$$f(x) = x + \tfrac{1}{2} \sin f(x).$$

(Consider continuous functions such that $|f(x)| \leqslant 2$.)

2. Extend 1.2.2 to the case where $T^k$ is a contraction mapping for some integer $k > 1$ (see 5.2.1).

3. If $\mathcal{M}$ is a compact non-empty metric space, $T$ maps $\mathcal{M}$ into $\mathcal{M}$ and $\rho(Tx, Ty) < \rho(x, y)$ for $x \neq y$, show that $T$ has a unique fixed point (see 5.2.3).

4. If $T$ is a contraction mapping of a Banach space $\mathcal{V}$ into itself, show that the equation $Tf - f = g$ has a unique solution $f$ for each $g$ in $\mathcal{V}$. Also show that $T - I$ and $(T - I)^{-1}$ are uniformly continuous (see 4.4.2).

# 2. *Fixed points in compact convex sets*

The principal results of this chapter are Brouwer's theorem (2.1.11), Schauder's theorem (2.3.7) and Tychonoff's theorem (2.3.8), all of which assert that every continuous mapping of a compact convex set into itself must have a fixed point. We end with an example showing that it is not enough for the set to be bounded, complete and convex.

## 2.1 The fixed point property

DEFINITION 2.1.1   A topological space $\mathcal{X}$ is said to possess the *fixed point property* if every continuous mapping of $\mathcal{X}$ into $\mathcal{X}$ has a fixed point.

It is often possible to decide that a set has *not* got the fixed point property, by finding a mapping without fixed points. (Consider, for instance, the real line or the unit circle.)

An elementary argument shows that the unit interval $[0, 1]$ has the fixed point property. A fairly simple argument (see §10.1) shows that the closed unit disc in the plane has the fixed point property. In all other important cases the fixed point property is rather hard to establish.

We observe first that the fixed point property is a topological property.

THEOREM 2.1.2   *If $\mathcal{X}$ is homeomorphic to $\mathcal{Y}$ and $\mathcal{X}$ has the fixed point property then $\mathcal{Y}$ has the fixed point property.*
   *Proof.* An exercise.   □

Using 2.1.2 and the results for the disc, one can show that various plane sets in amoeboid shapes have the fixed point property. But to deal with a spider's shape with a two-dimensional body and one-dimensional legs, or with a string of beads, one needs the next theorem.

DEFINITION 2.1.3   We say that $\mathscr{X}$ *is a retract of* $\mathscr{Y}$ if $\mathscr{X} \subset \mathscr{Y}$ and there exists a continuous mapping $r$ of $\mathscr{Y}$ into $\mathscr{X}$ such that $r = I$ on $\mathscr{X}$. (We then call $r$ a *retraction* mapping.)

EXAMPLE 2.1.4   *A closed convex non-empty subset* $\mathscr{X}$ *of* $E^n$ *or of a Hilbert space is a retract of any larger subset.*

*Sketch of proof.* The required retraction mapping is obtained by mapping each point onto the nearest point of $\mathscr{X}$. For details see Bourbaki (1955, 5.1.4). □

(The same result holds in Banach spaces but requires a different proof: see Dugundji (1958, theorem 10.2). For the case where $\mathscr{X}^0 \neq \varnothing$, see the proof of 4.2.4. below.)

THEOREM 2.1.5   *If* $\mathscr{Y}$ *has the fixed point property and* $\mathscr{X}$ *is a retract of* $\mathscr{Y}$ *then* $\mathscr{X}$ *has the fixed point property.*

*Proof.* Let $r$ be a retraction map of $\mathscr{Y}$ onto $\mathscr{X}$. If $T$ is any continuous map of $\mathscr{X}$ into $\mathscr{X}$ then $Tr$ is a continuous map of $\mathscr{Y}$ into $\mathscr{X}$. Since $Tr$ maps $\mathscr{Y}$ into $\mathscr{Y}$, there is a fixed point $w$, thus $Trw = w$. Clearly $w \in \mathscr{X}$ so that $rw = w$ and hence $Tw = w$. □

DEFINITION 2.1.6   A topological space $\mathscr{X}$ is *contractible* (to a point $x_0$ in $\mathscr{X}$) if there exists a continuous function $f(x, t)$ on $\mathscr{X} \times [0, 1]$ to $\mathscr{X}$ such that $f(x, 0) \equiv x$ and $f(x, 1) \equiv x_0$.

In order to obtain a fairly intuitive proof of Brouwer's theorem we will assume known the following facts about homology groups. (In §2.2, we discuss some proofs of Brouwer's theorem which do not require homology theory; in Chapter 10 we refer to some short proofs requiring more algebraic topology.) We write $S^n$ for the $n$-sphere and $B^n$ for the closed $n$-ball.

REMINDER 2.1.7   With each complex $\mathscr{X}$ in Euclidean space and each integer $n \geqslant 1$, we can associate a unique group $H_n(\mathscr{X})$ (the *nth homology group* with integral coefficients). Also

(i) $H_n(S^n) = Z$, the group of integers;
(ii) if $\mathscr{X}$ is contractible, then $H_n(\mathscr{X}) = \{e\}$, the trivial group.

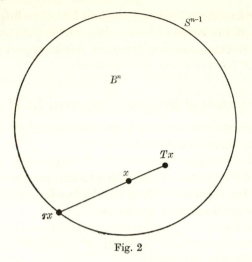

Fig. 2

**T H E O R E M   2.1.8**   *For $n \geqslant 0$, $S^n$ is not contractible.*

*Proof.* This result is obvious for $n = 0$ and follows from 2.1.7 for $n \geqslant 1$. $\square$

**L E M M A   2.1.9**   *If $\mathcal{Y}$ is contractible then any retract of $\mathcal{Y}$ is contractible.*

*Proof.* If $r$ retracts $\mathcal{Y}$ onto $\mathcal{X}$, and the function $f(x,t)$ contracts $\mathcal{Y}$ to a point $z \in \mathcal{Y}$, it is easily verified that $rf(x,t)$ contracts $\mathcal{X}$ to the point $rz \in \mathcal{X}$. $\square$

**T H E O R E M   2.1.10**   *For $n \geqslant 1$, $S^{n-1}$ is not a retract of $B^n$.*

*Proof.* $B^n$ is obviously contractible. By 2.1.8, $S^{n-1}$ is not. The result follows from 2.1.9. $\square$

**T H E O R E M   2.1.11** (*Brouwer*, 1910):

   (i)   *$B^n$ has the fixed point property.*

   (ii)  *Every compact convex non-empty subset $\mathcal{X}$ of $E^n$ has the fixed point property.*

   *Proof.*

   (i)   If there existed a map $T$ of $B^n$ into $B^n$ without fixed points, we could produce a retraction of $B^n$ onto $S^{n-1}$ as follows: for each $x$ in $B^n$, extend the line segment from $Tx$ through $x$ to the point of intersection with $S^{n-1}$; call this point $rx$ (see figure 2). Such a retraction is impossible by 2.1.10.

(ii) For $k$ sufficiently large, the ball $kB^n$ of radius $k$ contains $\mathscr{X}$. By 2.1.4, $\mathscr{X}$ is a retract of $kB^n$. Since $kB^n$ is homeomorphic to $B^n$, 2.1.2 shows that $kB^n$ has the fixed point property and 2.1.5 shows that $\mathscr{X}$ has the fixed point property. $\square$

## 2.2   Other proofs of Brouwer's theorem 2.1.11

Bohl (1904) proved a result equivalent to 2.1.10 but apparently did not go on to obtain 2.1.11.

Proofs of Brouwer's theorem 2.1.11 depending on various definitions of the degree of a mapping (rotation of a vector field) were given by Brouwer (1910) by Alexander (1922), and by various later authors; see Chapter 10.

Anyone preferring a proof by classical methods (calculus and determinants) should refer to Birkhoff and Kellogg (1922), or Dunford and Schwartz (1958).

The most direct method of proof is by simplicial subdivision of an $n$-simplex. The proof given in Knaster, Kuratowski and Mazurkiewicz (1929) and also presented in Kuratowski (1933) and Graves (1946) involves the following preliminary result.

NOTATION   We write co$(X)$ for the convex cover of a set $X$.

THEOREM 2.2.1   *(K., K. and M.) Given $n+1$ closed subsets $A(0), A(1), ..., A(n)$ of an $n$-simplex*

$$S = \mathrm{co}(p(0), p(1), ..., p(n)),$$

*such that each face* co$(p(i_0), ..., p(i_k))$ *satisfies*

$$\mathrm{co}(p(i_0), ..., p(i_k)) \subset A(i_0) \cup ... \cup \mathrm{A}(i_k),$$

*we must have $A(0) \cap ... \cap A(n) \neq \varnothing$.*

*Sketch of proof.* Using a combinatorial argument (Sperner's lemma) it is shown that an arbitrarily fine simplicial subdivision of $S$ contains a simplex with vertices in all the $A_i$. A limiting process then gives a point lying in all the $A_i$.

From 2.2.1 we derive Brouwer's theorem for the $n$-simplex $S = \mathrm{co}(p(0), p(1), ..., p(n))$ as follows. Express each $x$ in $S$ uniquely in the form $x = \Sigma x_i p(i)$. (Here $x_i \geqslant 0$ and $\Sigma x_i = 1$.) If we take any continuous mapping $T$ of $S$ into $S$ and set

$$A_i = \{x : (Tx)_i \leqslant x_i\},$$

we see that the $A_i$ satisfy the conditions of 2.2.1. This gives a point $x$ such that $(Tx)_i \leqslant x_i$ for all $i$. Since $\Sigma(Tx)_i = \Sigma x_i = 1$ we must have $(Tx)_i = x_i$ for all $i$, so that $x$ is a fixed point for $T$.

Perhaps the most intuitive proof of 2.1.11 is obtained by using the elementary proof in Hirsch (1963) of 2.1.10; for the details see Maunder (1970). ☐

## 2.3   Extensions to infinite-dimensional spaces

Most applications of topological theorems to analysis involve infinite-dimensional spaces of functions or sequences. (For an exception see §6.4.) The usual procedure is to extend a theorem from the finite-dimensional case to the infinite-dimensional case; we approximate infinite-dimensional sets or mappings by finite-dimensional sets or mappings. In this section we carry out this extension process from Brouwer's theorem to its analogue (Schauder's theorem) for an infinite-dimensional Banach space. We will use the following approximation lemma.

LEMMA 2.3.1   *Let $\mathscr{Y}$ be a compact metric space. For each $\epsilon > 0$, let $P_\epsilon$ be a continuous mapping of $\mathscr{Y}$ into $\mathscr{Y}$ such that $\rho(P_\epsilon x, x) < \epsilon \, (\forall x)$. Suppose that each set $P_\epsilon \mathscr{Y}$ has the fixed point property. Then $\mathscr{Y}$ has the fixed point property.*

*Proof.* Consider a continuous mapping $T$ of $\mathscr{Y}$ into $\mathscr{Y}$. Since $P_\epsilon T$ maps $P_\epsilon \mathscr{Y}$ into itself, there is a fixed point $x_\epsilon$; that is, $P_\epsilon T x_\epsilon = x_\epsilon$.

Thus $\rho(x_\epsilon, T x_\epsilon) = \rho(P_\epsilon T x_\epsilon, T x_\epsilon) < \epsilon$. By 1.1.3, $T$ has a fixed point in $\mathscr{Y}$. ☐

We show now that it is sufficient to consider a particular infinite-dimensional space, the Hilbert cube.

DEFINITION 2.3.2   The *Hilbert cube* $\mathscr{H}_0$ is the subset of $l^2$ consisting of points $a = (a_1, a_2, \ldots)$ such that $|a_r| \leqslant r^{-1}$ for all $r$.

THEOREM 2.3.3   *Every compact convex subset $\mathscr{K}$ of a Banach space $\mathscr{B}$ is homeomorphic, under a linear mapping, to a compact convex subset of $\mathscr{H}_0$.*

*Proof.* We assume, without loss of generality, that $\mathscr{K}$ is a subset of the unit ball in $\mathscr{B}$. Since $\mathscr{K}$ and span $(\mathscr{K})$ are separable we can choose a sequence $(x_n)$ dense in span $(\mathscr{K})$. For

$$n = 1, 2, \ldots$$

choose $f_n$ in the dual space $\mathscr{B}^*$ such that

$$f_n(x_n) = \|x_n\|/n, \quad \|f_n\| = 1/n.$$

Then the mapping

$$F: x \to (f_1(x), \, f_2(x), \, \ldots, \, f_n(x), \, \ldots)$$

clearly maps $\mathscr{K}$ into $\mathscr{H}_0$. We can see that $F$ is a bounded linear operator on $\mathscr{B}$ to $l^2$. $F$ is one–one on span $(\mathscr{K})$ since if $x \neq y$ in span $(\mathscr{K})$ we have

$$|f_n(x) - f_n(y)| \geqslant |f_n(x_n)| - |f_n(x - y - x_n)|$$
$$\geqslant \|x_n\|/n - \|(x - y) - x_n\|/n > 0,$$

if $x_n$ is sufficiently close to $x - y$. Thus $F$ is a homeomorphism on $\mathscr{K}$ to $F(\mathscr{K})$, since $F$ is one–one and continuous on the compact set $\mathscr{K}$.

We now conclude that $F\mathscr{K}$ is compact and convex since a linear homeomorphism preserves these properties. $\square$

**DEFINITION 2.3.4**  $P_n$ is the projection of $l^2$ onto an $n$-dimensional subspace given by

$$P_n(x_1, x_2, \ldots) = (x_1, x_2, \ldots, x_n, 0, 0, \ldots).$$

**THEOREM 2.3.5**  *The Hilbert cube* $\mathscr{H}_0$ *has the fixed point property.*

*Proof.* We observe that for $n$ sufficiently large,

$$\|P_n a - a\| \leqslant \left( \sum_{n+1}^{\infty} r^{-2} \right)^{\frac{1}{2}} < \epsilon \tag{1}$$

for all $a$ in $\mathscr{H}_0$. Since $P_n \mathscr{H}_0$ is compact, this shows that $\mathscr{H}_0$ is compact. Since $P_n \mathscr{H}_0$ can be regarded as a compact convex subset of $R^n$, Brouwer's theorem 2.1.11 gives the fixed point property for $P_n \mathscr{H}_0$. Thus 2.3.1 shows that $\mathscr{H}_0$ has the fixed point property. $\square$

**THEOREM 2.3.6**  *Any non-empty compact convex subset* $\mathscr{X}$ *of* $\mathscr{H}_0$ *has the fixed point property.*

*Proof.* By 2.1.4, $\mathscr{X}$ is a retract of $\mathscr{H}_0$; by 2.1.5, $\mathscr{X}$ has the fixed point property. $\square$

THEOREM 2.3.7 (*Schauder*, 1930)  *Any compact convex non-empty subset $\mathscr{Y}$ of a normed space has the fixed point property.*

*Proof.* By 2.3.3, $\mathscr{Y}$ is homeomorphic to a compact convex subset $\mathscr{X}$ of $\mathscr{H}_0$; by 2.3.6, $\mathscr{X}$ has the fixed point property; thus 2.1.2 gives the result. □

REMARK  Schauder's original proof of 2.3.7 involves approximating $\mathscr{Y}$ by the convex cover $\mathrm{co}(\mathscr{X})$ of a finite subset $\mathscr{X}$ of $\mathscr{Y}$. A simplicial subdivision of $\mathrm{co}(\mathscr{X})$ is then made, and any mapping $T$ of $\mathscr{Y}$ into $\mathscr{Y}$ is approximated by a simplicial mapping of $\mathrm{co}(\mathscr{X})$ into $\mathrm{co}(\mathscr{X})$. Fixed points for these simplicial mappings exist by 2.1.11 and yield, by compactness, a fixed point for $T$. Modern versions of this proof avoid simplicial subdivision; see our proof of Schauder's second theorem (4.1.1).

The simplicial subdivision method was used in the original proof of the following theorem.

THEOREM 2.3.8 (*Tychonoff*, 1935)  *Any compact convex non-empty subset of a locally convex space has the fixed point property.*

We will sketch a proof of a more general result in §4.5.

We mention some other proofs of the theorems of Schauder and Tychonoff.

Before Schauder gave his general theorem, proofs for particular function spaces were given by Birkhoff and Kellogg (1922), by interpolation between a finite number of function values in the cases $C$ and $C^{(k)}$, and by using Fourier series in the $L^2$ case; see also Schauder (1927).

A proof of Tychonoff's theorem using the fixed point property for the Hilbert cube is given in Dunford and Schwartz (1958); a proof directly from 2.2.1 was found by Ky Fan (1961).

## 2.4  Kakutani's example:  *A fixed-point-free mapping of the unit ball in Hilbert space.*

(This example shows that in Schauder's theorem 2.3.7 the condition '$\mathscr{Y}$ is compact' cannot be replaced by '$\mathscr{Y}$ is bounded and closed'.)

We consider $l^2$ as $l^2(Z)$ with the natural basis consisting of sequences $y_n = (\ldots, 0, 0, 1, 0, 0, \ldots)$ with the 1 in position $n$. For $x$ in $l^2$ we can write

$$x = (\ldots, x_{-1}, x_0, x_1, x_2, \ldots) = \Sigma x_n y_n.$$

We write $U$ for the right-shift operator:

$$Ux = \Sigma x_n y_{n+1}.$$

LEMMA 2.4.1   *The vector $x - Ux$ is a multiple of $y_0$ only if $x = 0$.*

*Proof.* The relation

$$x - Ux = \Sigma(x_n - x_{n-1}) y_n = cy_0$$

requires that $x_n = x_0$ for all $n > 0$, and $x_n = x_{-1}$ for all $n < 0$; for a member of $l^2$, this is only possible if $x_0 = x_{-1} = 0$.  $\square$

THEOREM 2.4.2   *The unit ball $\mathscr{B}$ in $l^2$ lacks the fixed point property.*

(For extensions see Klee (1955), Dugundji (1951).)

*Proof.* We will define the mapping by

$$Tx = (1 - \|x\|)y_0 + Ux.$$

$T$ is continuous, and maps $\mathscr{B}$ into $\mathscr{B}$ since if $\|x\| \leqslant 1$ we have

$$\|Tx\| \leqslant (1 - \|x\|) \|y_0\| + \|Ux\|$$
$$= (1 - \|x\|) + \|x\| = 1.$$

Finally, $T$ has no fixed point, since if

$$x = Tx = (1 - \|x\|)y_0 + Ux$$

then $x - Ux = (1 - \|x\|)y_0$ which is clearly impossible if $x = 0$ and is impossible by the lemma if $x \neq 0$.  $\square$

REMARK 2.4.3   We can obtain a slightly stronger result by considering

$$T_c x = c(1 - \|x\|)y_0 + Ux \quad (0 < c < 1).$$

For each fixed $c$ this gives a homeomorphism of $\mathscr{B}$ onto $\mathscr{B}$, without fixed points. We also have

$$\|T_c x - T_c y\| \leqslant (1 + c) \|x - y\|.$$

For discussion of $T_{\frac{1}{2}}$ see Kakutani (1943) or Cronin (1964).

From the theorem we obtain

COROLLARY 2.4.4   *The unit sphere $S^\infty$ in $l^2$ is a retract of the unit ball in $l^2$.*

*Proof.* See the argument of 2.1.11 (i).  $\square$

## Exercises

1. Show graphically that the unit interval $[0, 1]$ has the fixed point property.

2. Let $S$ be the unit sphere and $B$ the closed unit ball in an infinite-dimensional Hilbert space $H$. Then

(i) there is a continuous mapping of $B$ into $B$ with no fixed points (use 2.4.2);

(ii) $S$ is a retract of $B$ (use the argument of 2.1.11 (i)).

(Both results hold in any Banach space: see Dugundji (1951).)

3. Show that each homeomorphism of a set $M$ into $M$ may have a fixed point, although $M$ lacks the fixed point property. (Consider a closed disc with a one-dimensional 'handle'.)

4. Unsolved problem: can Tychonoff's theorem be extended to arbitrary topological vector spaces?

5. (*Fort*, 1954) Let $B^0$ be an open ball in $R^n$ and $T$ a continuous mapping of $B^0$ into $B^0$. Show that for each $\epsilon > 0$

(i) there is a retraction $r$ of $B^0$ onto a closed convex subset of $B^0$, such that $\|rx - x\| < \epsilon$ $(\forall x \in B^0)$;

(ii) $T$ has an $\epsilon$-fixed point.

# 3. Which sets have the fixed point property?

## 3.1 Compact contractible sets

As a rough guide (for purposes of functional analysis) we expect that a set with the fixed point property should be compact and contractible. In fact, if a set lacks one of these properties, we can usually produce a mapping without fixed points as follows.

If a subset of $R^n$ is not compact we can usually produce a fixed-point-free mapping by moving all points towards a missing limit point, or 'towards infinity' (in some direction). Thus we see that sets such as an open interval or open ball, or a half-line or subspace, lack the fixed point property.

The unit ball in $l^2$, which is bounded and closed but not compact, lacks the fixed point property, since we saw (§2.4) that there is a fixed-point-free mapping in this set. Klee (1955) shows that any convex non-compact subset of a normed space lacks the fixed point property.

In simple cases, where a subset of $R^n$ is not contractible, it usually has some sort of a hole in the middle. In these cases we can rotate the set about the hole or reflect it through the hole – thus we see that sets such as a circle, sphere, Klein bottle, torus or Möbius strip lack the fixed point property.

THEOREM 3.1.1 *The sets mentioned in the three previous paragraphs lack the fixed point property.*

*Proof.* In each case we can construct a mapping of the set into itself, without fixed points, in the way suggested above. □

In §3.2 we will give some examples of sets which fail to be compact or contractible, but which have the fixed point property.

We now give some positive results.

THEOREM 3.1.2 (*Lefschetz*) *If $\mathscr{X}$ is a compact locally contractible metric space, all of whose homology groups are trivial, then $\mathscr{X}$ has the fixed point property.*

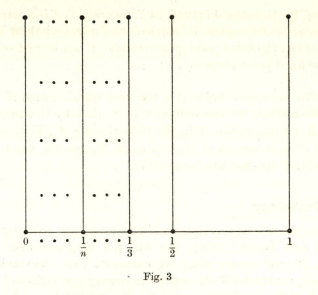

Fig. 3

*Proof.* See Lefschetz (1930, p. 359). A slightly different version of this theorem is given in Lefschetz (1942). See also 10.44. □

COROLLARY 3.1.3   *If $\mathscr{X}$ is a compact contractible and locally contractible metric space then $\mathscr{X}$ has the fixed point property.*

*Proof.* Immediate from 3.1.2. (A simple direct proof would be desirable.) □

We mention an example not covered by the theorem: the set shown, consisting of an infinite number of vertical closed segments based on a horizontal closed segment, has the fixed point property (figure 3).

A modern proof of 3.1.3 is given by Spanier (1966) under the assumption:        $\mathscr{X}$ *is a complex in $R^n$.*

Spanier's argument, following that of Lefschetz, is 'internal'; the action of a mapping of $\mathscr{X}$ is studied in terms of its effect on the homology groups of $\mathscr{X}$. We will now obtain a more general result than that of Spanier, by an 'external' argument.

THEOREM 3.1.4   *If $\mathscr{X}$ is a compact, contractible and locally contractible subset of $R^n$ then $\mathscr{X}$ has the fixed point property.*

*Proof*. By Kuratowski (1968, 54.7, theorem 6) $\mathscr{X}$ is a retract of any larger metric space, in particular of a large ball $\mathscr{B}$ in $R^n$. Since $\mathscr{B}$ has the fixed point property and $\mathscr{X}$ is a retract of $\mathscr{B}$, $\mathscr{X}$ has the fixed point property. □

A similar argument holds if a compact metric space $\mathscr{X}$ is an absolute retract. We can embed $\mathscr{X}$ in $l^\infty$, then by the argument of 2.3.3, we can embed $\mathscr{X}$ in the Hilbert cube $\mathscr{H}_0$. Thus we can regard $\mathscr{X}$ as a retract of $\mathscr{H}_0$ and since $\mathscr{H}_0$ has the fixed point property, $\mathscr{X}$ also has this property.

## 3.2  Pathology

Many interesting examples and references are given by Bing (1969) and Fadell (1970). We will not compete with these works. We will merely show by examples that the conditions compact, contractible are neither necessary nor sufficient for a set to have the fixed point property. It is enough to consider sets in $R^3$.

THEOREM 3.2.1 (*Kinoshita*)  *There exists a compact contractible subset of $R^3$ which lacks the fixed point property.*
*Proof*. See Kinoshita (1953) for the very elegant example which solved this question – which had been open for at least twenty years. The set in question is the union of a horizontal closed disc, a vertical cylinder of unit height based on the edge of the disc, and a vertical sheet of unit height and infinite length which spirals out from the axis of the cylinder, approaching closer and closer to the cylinder. (See figure 4). □

THEOREM 3.2.2  *The projective plane, or any (real) projective space of even dimension, has the fixed point property.*
*Proof*. See for instance Whittlesey (1963, Corollary 17). □

By 3.2.2, contractibility is not necessary for the fixed point property. Examples showing that compactness is not necessary are inevitably more pathological. We give a theorem which provides us with examples (for an earlier version see Smart (1967)).

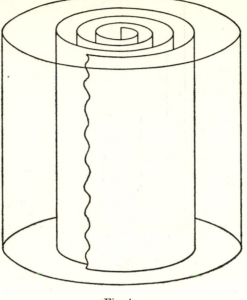

Fig. 4

THEOREM 3.2.3   *The sets shown in figures* 3, 5 *and* 6 *have the fixed point property. The set in figure* 5 *is not compact; that in figure* 6 *is neither compact nor contractible.*

We can describe the sets as follows. In each case the set has the form $\mathscr{X} = \mathscr{X}_0 \cup \bigcup_1^\infty \mathscr{X}_n$ where each $\mathscr{X}_n$ is homeomorphic to a closed line segment and for $n > 0$, $\mathscr{X}_n$ is attached to $\mathscr{X}_0$ at $p_n$, an endpoint of $\mathscr{X}_n$. In figure 5, $\mathscr{X}_0 = [0, 1]$, $p_n \to 0$ and $\mathscr{X}_n$ is a segment of height $n$ over $p_n$. In figure 6, $\mathscr{X}_0$ is a semicircle in the lower half plane joining $(0,0)$ to $(1, 0)$ and for $n \geqslant 1$, $\mathscr{X}_n$ is the line segment joining $p_n = (1, 0)$ to the point $(0, n^{-1})$.

*Proof.* First observe:

  *any arc joining a point in* $\mathscr{X}_i$ *to a point in* $\mathscr{X} - \mathscr{X}_i$ *must pass through* $p_i$.                                              (1)

Let $T$ be any continuous mapping of $\mathscr{X}$ into $\mathscr{X}$. Consider three cases.

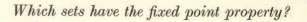

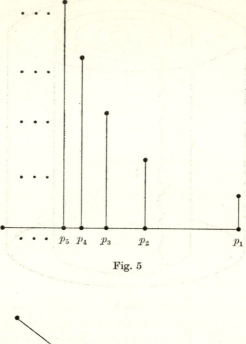

Fig. 5

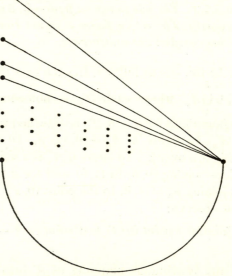

Fig. 6

*Case* 1. For some $i \geqslant 1$, $Tp_i = p_i$. Then $p_i$ is a fixed point.

*Case* 2. For some $i \geqslant 1$, $Tp_i \in \mathscr{X}_i - \{p_i\}$. Define a mapping $R$ of $\mathscr{X}$ onto $\mathscr{X}_i$ by

$$Rx = \begin{cases} x & (x \in \mathscr{X}_i), \\ p_i & (x \notin \mathscr{X}_i). \end{cases}$$

Then by (1),

$Rx(t)\,(0 \leqslant t \leqslant 1)$ *will be an arc in* $\mathscr{X}_i$ *if* $x(t)\,(0 \leqslant t \leqslant 1)$ *is an arc in* $\mathscr{X}$. $\hfill (2)$

If we put $Sx = RTx$, $S$ gives a mapping of $\mathscr{X}_i$ into $\mathscr{X}_i$; we will show that $S$ is continuous. If $x_n \to x$ in $\mathscr{X}_i$, we can assume that $x_n \to x$ along an arc in $\mathscr{X}_i$; thus $Tx_n \to Tx$ along an arc in $\mathscr{X}$ so that by (2), $RTx_n \to RTx$, that is, $Sx_n \to Sx$. Thus $S$ is continuous and so has a fixed point in $\mathscr{X}_i$. Since $z \neq p_i$ we have $Tz = Sz = z$.

*Case* 3. For all $i \geqslant 1$, $Tp_i \notin \mathscr{X}_i$. Define a mapping $R$ of $\mathscr{X}$ onto $\mathscr{X}_0$ by

$$Rx = \begin{cases} x & (x \in \mathscr{X}_0), \\ p_i & (x \in \mathscr{X}_i, i \geqslant 1). \end{cases}$$

As in case 2 we see that $Sx = RTx$ is continuous. Thus $S$ has a fixed point $z$. Since $z$ cannot be a $p_i$ we have $Tz = Sz = z$.

Thus in each case $T$ has a fixed point. $\hfill \square$

We can extend 3.2.3, by the same argument, to other cases where $\mathscr{X}$ is a metric space of the form $\mathscr{X} = \cup\, \mathscr{X}_i$ and

(i)  each $\mathscr{X}_i$ is locally arcwise connected and has the fixed point property;

(ii)  each $\mathscr{X}_i\,(i \neq 0)$ is joined to $\mathscr{X}_0$ by a single point $p_i$ and any arc from a point in $\mathscr{X}_i$ to a point in $\mathscr{X} - \mathscr{X}_i$ must pass through $p_i$;

(iii)  the number of $\mathscr{X}_i$ is finite, countable or uncountable.

In particular we obtain the following example by suitably stretching and bending the segments in figure 5. (We arrange for $\bigcup_0^n \mathscr{X}_i$ to be $n^{-1}$-dense in the ball of radius $n$.)

**T H E O R E M  3.2.4** (*The Spaghetti Set*)  *For* $k \geqslant 2$, *there exists an everywhere dense subset of* $R^k$ *which has the fixed point property.*

**Exercises**

1. Consider sets in the plane in the shape of the letters $A$, $B$, $C$, $D$, $E$. Which of these sets have the fixed point property?

2. Give an ad hoc proof that the set shown in figure 3 has the fixed point property.

3. Show that any non-compact convex subset of $R^n$ lacks the fixed point property.

4. If $X$ is any topological space then $X \times S^n$ lacks the fixed point property.

5. Can we extend 3.2.4 to the case $k = 1$?

# 4. *Extensions of Schauder's theorem*

We give first a more general form of Schauder's theorem, and then other results which follow by quite simple arguments. Each of these theorems trivially implies Schauder's theorem.

For many applications of fixed point theorems, the forms given in this chapter are the easiest to use.

## 4.1 Schauder's second theorem

THEOREM 4.1.1 *Let $\mathcal{M}$ be a non-empty convex subset of a normed space $\mathcal{B}$. Let $T$ be a continuous mapping of $\mathcal{M}$ into a compact set $\mathcal{K} \subset \mathcal{M}$. Then $T$ has a fixed point.*

Schauder (1930) proved this result in the case where $\mathcal{B}$ is complete and $\mathcal{M}$ closed. The argument is, to consider $T$ as a mapping of the compact convex set $\overline{\mathrm{co}}(T\mathcal{M})$ into itself, and use 2.3.7. To prove the general result, we give an argument directly from 2.1.11.

DEFINITION 4.1.2 *Let $T$ map a set $\mathcal{S}$ into a topological space $\mathcal{X}$. If $T\mathcal{S}$ is contained in a compact subset of $\mathcal{X}$, we say that $T$ is compact.*

NOTATION We write $\mathrm{co}(\mathcal{X})$ for the smallest convex set containing $\mathcal{X}$, and $\overline{\mathrm{co}}(\mathcal{X})$ for the closure of $\mathrm{co}(\mathcal{X})$.

LEMMA 4.1.3 (*'Schauder's Projection'*) *If $\mathcal{K}$ is a compact subset of a normed space $\mathcal{V}$ and $\epsilon > 0$, there is a finite subset $\mathcal{X}$ of $\mathcal{K}$ and a continous mapping $P$ of $\mathcal{K}$ into $\mathrm{co}(\mathcal{X})$ such that*

$$\|Px - x\| < \epsilon \quad (x \in \mathcal{K}).$$

*Proof.* Choose $x_1, \ldots, x_n$ in $\mathcal{K}$ such that the sets $N(x_i, \epsilon)$ with $1 \leqslant i \leqslant n$ cover $\mathcal{K}$. Put $\mathcal{X} = \{x_1, \ldots, x_n\}$. For $1 \leqslant i \leqslant n$ put

$$f_i(x) = \max{(0, \epsilon - \|x - x_i\|)}.$$

[ 25 ]

Then $f_i(x) \neq 0$ if and only if $x \in N(x_i, \epsilon)$. Thus at each $x$ in $\mathscr{K}$, some $f_i(x) \neq 0$. Now put

$$Px = \Sigma f_i(x) x_i / \Sigma f_i(x) \quad (x \in \mathscr{K}).$$

Clearly $P$ is continuous. Also, since $Px$ is a convex combination of those points $x_i$ which lie in $N(x, \epsilon)$, we have $Px \in N(x, \epsilon)$. □

*Proof of* 4.1.1. For $n = 1, 2, \ldots$ consider $P_n T$ where $P_n$ is the mapping given by 4.1.3 with $\epsilon = 1/n$. Since $\mathscr{X} \subset \mathscr{K} \subset \mathscr{M}$ we have $\mathrm{co}(\mathscr{X}) \subset \mathscr{M}$. Thus $P_n T$ gives a continuous mapping of the finite-dimensional compact convex set $\mathrm{co}(\mathscr{X})$ into itself. A fixed point $x_n$ exists by Brouwer's theorem 2.1.11. From $P_n T x_n = x_n$ we get $\|T x_n - x_n\| < 1/n$. By 1.1.3, $T$ has a fixed point. □

The following special case of 4.1.1 is useful for applications.

COROLLARY 4.1.4 *Let $T$ be a compact continuous mapping of a normed space $\mathscr{B}$ into $\mathscr{B}$. Then $T$ has a fixed point.*

Browder (1959) extends 4.1.1 and 4.1.4 to the case where some power $T^n (n > 1)$ is compact.

For finite-dimensional spaces the above results become:

THEOREM 4.1.5 (a) *Any continuous mapping of a convex subset $\mathscr{M}$ of $R^n$ into a bounded closed set inside $\mathscr{M}$ has a fixed point.*

(b) *Any continuous mapping of $R^n$ into a bounded subset of $R^n$ has a fixed point.*

## 4.2　Rothe's theorem

We will consider cases where a set $\mathscr{M}$ is not mapped into itself; however, the boundary of $\mathscr{M}$ must be mapped into $\mathscr{M}$.

NOTATION We write $\mathscr{M}^0$ for the interior of a set $\mathscr{M}$ and $\partial \mathscr{M}$ for the boundary of $\mathscr{M}$.

LEMMA 4.2.1 *Let $\mathscr{M}$ be the closed ball of radius $n$ in a normed space $\mathscr{B}$. The radial retraction onto $\mathscr{M}$ is defined by*

$$rx = \begin{cases} x & \text{if } x \in \mathscr{M}, \\ nx/\|x\| & \text{if } x \notin \mathscr{M}. \end{cases}$$

*Then*

    (i)   *r is a continuous retraction of $\mathscr{B}$ onto $\mathscr{M}$,*

    (ii)  *if $rx \in \mathscr{M}^0$ then $rx = x$,*

    (iii) *if $x \notin \mathscr{M}$ then $rx \in \partial \mathscr{M}$.*

*Proof.* Obvious. $\square$

**LEMMA 4.2.2**   *Let $T: \mathscr{M} \to \mathscr{N}$ be compact and let $r: \mathscr{N} \to \mathscr{P}$ be continuous. Then $rT$ is compact.*

*Proof.* Obvious. $\square$

**THEOREM 4.2.3** *(Rothe, 1937)*   *Let $\mathscr{B}$ be a normed space, $\mathscr{M}$ the closed unit ball in $\mathscr{B}$ and $\partial \mathscr{M}$ the unit sphere in $\mathscr{B}$. Let $T$ be a continuous compact mapping of $\mathscr{M}$ into $\mathscr{B}$ such that $T(\partial \mathscr{M}) \subset \mathscr{M}$. Then $T$ has a fixed point.*

*Proof.* Let $r$ be the radial retraction onto $\mathscr{M}$. Then $rT$ is compact by 4.2.2 so has a fixed point $y$ by 4.1.1;

$$rTy = y.$$

If $y \in \partial \mathscr{M}$ then $Ty \in \mathscr{M}$ so that

$$y = rTy = Ty.$$

These equations are still true, by 4.2.1 (ii), if $y = rTy$ is in $\mathscr{M}^0$. $\square$

We can generalise the previous result.

**THEOREM 4.2.4**   *Theorem 4.2.3 is still true if we allow $\mathscr{M}$ to be any closed convex subset of $\mathscr{B}$ with $\partial \mathscr{M}$ the boundary of $\mathscr{M}$.*

For the proof we require a lemma.

**LEMMA 4.2.5**   *Let $\mathscr{M}$ be a closed convex subset of a normed space $\mathscr{B}$ such that $0 \in \mathscr{M}^0$. Then the Minkowski functional*

$$g(x) = \inf\{c: x \in c\mathscr{M}\}$$

*is a continuous real function on $\mathscr{B}$ such that*

    (i)    $g(cx) = cg(x)$ *for* $c \geqslant 0$,

    (ii)   $g(x+y) \leqslant g(x) + g(y)$,

    (iii)  $0 \leqslant g(x) < 1$ *if* $x \in \mathscr{M}^0$,

    (iv)  $g(x) > 1$ *if* $x \notin \mathscr{M}$,

    (v)   $g(x) = 1$ *if* $x \in \partial \mathscr{M}$.

*Proof.* We remark that continuity follows from (ii) and continuity at 0; the other properties are easily proved. □

*Proof of theorem* 4.2.4 (due essentially to Potter (1973)). The result is trivial if $\mathcal{M}^0 = \varnothing$, so assume without loss of generality that $0 \in \mathcal{M}^0$. We define the radial retraction $r$ of $\mathcal{B}$ onto $\mathcal{M}$ by

4.2.6 $$rx = x/\max(1, g(x)).$$

By 4.2.5, $r$ has the properties (i) to (iii) of 4.2.1. Thus the proof of 4.2.3 can be followed. □

REMARK In 4.2.3 and 4.2.4. we cannot replace the assumption '$T$ *is compact on* $\mathcal{M}$' by the weaker assumption '$T$ *is compact on* $\partial\mathcal{M}$'. For example, consider the ball $\mathcal{M}$ of radius 2 in $l^2$ and the mapping $T$ of $\mathcal{M}$ into $\mathcal{M}$ defined by

$Tx$ is given by Kakutani's mapping for $\|x\| \leqslant 1$

$Tx = (2 - \|x\|)\,T(x/\|x\|)$ for $1 \leqslant \|x\| \leqslant 2$.

## 4.3 Continuation theorems

Let $\mathcal{M}$ be a region (that is, a connected open set) in a normed space $\mathcal{B}$. A number of theorems are concerned with a family of mappings $U_t$ ($0 \leqslant t \leqslant 1$) of $\mathcal{M}$ into $\mathcal{B}$ such that $U_t$ has no fixed points on the boundary $\partial\mathcal{M}$. This means that as $t$ changes, fixed points cannot 'escape' from $\mathcal{M}$ through $\partial\mathcal{M}$. Thus if $U_0$ satisfies suitable conditions (which ensure a fixed point for $U_0$) we expect that $U_1$ must have a fixed point; the theorems assert that this is so.

DEFINITION 4.3.1 Let $U_0$ and $U_1$ be mappings of a set $\mathcal{N}$ into $\mathcal{B}$. We say that $U_0$ *is fp-homotopic to* $U_1$ *on* $\mathcal{N}$ if there exists a family of mappings $U_t$ ($0 \leqslant t \leqslant 1$) of $\mathcal{N}$ into $\mathcal{B}$ such that
  (i) $U_t(x) = U(x, t)$ is continuous on $\mathcal{N} \times [0, 1]$,
  (ii) $U(\mathcal{N} \times [0, 1])$ is contained in a compact subset of $\mathcal{B}$,
  (iii) $U_t x \neq x$ for $x \in \partial\mathcal{N}$.

The continuation theorems have the following general form.

*If*      (*a*) *a condition on* $\mathcal{M}$,

         (*b*) *a condition on* $U_0$,

*and*    (*c*) $U_1$ *is fp-homotopic to* $U_0$ *on* $\partial\mathcal{M}$,     (G)

*then*        $U_1$ *has a fixed point.*

The first continuation theorems applicable to non-linear problems were due to Leray and Schauder (1934); we believe that when people mention 'the Leray–Schauder theorem' they usually mean our 10.3.10. This result is the most famous and most general result of the form (G). The condition on $U_0$ is that $\deg(I - U_0) \neq 0$; thus the theorem cannot be stated or applied without a knowledge of degree theory.

Various attempts have been made to replace the Leray–Schauder theorem by theorems in which the degree is not used. These theorems use conditions on $U_0$ and $\mathcal{M}$ which are less general but more easily established in applications. The most useful result is that of Schaefer ((1955); see also 4.3.2) which can be restated as a theorem of the form (G) using the strong simplifications $U_0 = 0$, $U_t = tU_1$, $\mathcal{M}$ is a ball in a $\mathcal{B}$-space. Browder ((1966, lemme 24) removes the restriction $U_t = tU_1$; his argument is adapted by Potter (1973) to the more general case where $U_0(\partial\mathcal{M}) \subset \mathcal{M}$ and $\mathcal{M}$ is convex, which we will give in 4.3.3. Granas (1961, 1962) considers a general region $\mathcal{M}$ and the initial condition: $V = U_{0|\partial\mathcal{M}}$ is essential on $\partial\mathcal{M}$ (that is, each continuous compact extension of $V$, to a mapping of $\mathcal{M}$ into $\mathcal{B}$, has a fixed point). This result is as general as the Leray–Schauder theorem, but the condition '$V$ is essential' may be as hard to verify as the condition '$\deg(I - U_0) \neq 0$'. Neither of these conditions can be generalized. In Smart (1973$a$), Granas's result is proved for the case where $\mathcal{M}$ is convex, by the method of Potter. Browder (1970) uses the initial condition: $I - U_0$ is a homeomorphism onto a neighbourhood of 0.

In the original work of Leray and Schauder the fixed point $P_t$ obtained for $U_t$ is a continuous function of $t$ on $[0, 1]$. We will not consider results of this type (but see for instance Browder (1960$a$)).

THEOREM 4.3.2 (*Schaefer*) *Let $\mathcal{B}$ be a normed space, $T$ a continuous mapping of $\mathcal{B}$ into $\mathcal{B}$ which is compact on each bounded subset $\mathcal{X}$ of $\mathcal{B}$. Then either*
  (i) *the equation $x = \lambda Tx$ has a solution for $\lambda = 1$, or*
  (ii) *the set of all such solutions $x$, for $0 < \lambda < 1$, is unbounded.*
  *Proof.* Consider the radial retraction $r$ onto the ball $\mathcal{M}$ of

radius $n$ in $\mathscr{B}$ (see 4.2.1). Then $rT$ has a fixed point $x$ in $\mathscr{M}$ by Schauder's theorem 4.1.3. Either (i) $\|Tx\| \leqslant n$, in which case $Tx = rTx = x$, or (ii) $\|Tx\| > n$, in which case $\|x\| = \|rTx\| = n$, so that

$$x = rTx = (n/\|Tx\|)\,Tx = \lambda Tx \quad \text{with } 0 < \lambda < 1.$$

Thus either for some integer $n$ we obtain a solution of $Tx = x$, or for each $n$ we obtain an eigenvector of norm $n$ for some eigenvalue in $(0, 1)$; in the second case the set of such eigenvectors is unbounded. $\square$

**T H E O R E M** 4.3.3 (*Browder–Potter*)   *Let $\mathscr{M}$ be a closed convex subset of a normed space $\mathscr{B}$. Let $U(x, t)$ be a continuous mapping of $\mathscr{M} \times [0, 1]$ into a compact subset of $\mathscr{B}$ such that*

*(A)* $U_0(\partial\mathscr{M}) \subset \mathscr{M}$,

*(B)* *for* $0 \leqslant t \leqslant 1$, $U_t$ *has no fixed point on* $\partial\mathscr{M}$ *(where*

$$U_t(x) = U(x, t)).$$

*Then*          $U_1$ *has a fixed point in* $\mathscr{M}$.

*Proof.* Without loss of generality we assume that $0 \in \mathscr{M}^0$. We assume that $U_1$ has no fixed point and obtain a contradiction. For $\epsilon > 0$ we define a mapping $S$ of $\mathscr{M}$ into $\mathscr{B}$ by

$$Sx = U_1(x/(1-\epsilon)) \qquad \text{if } g(x) \leqslant 1 - \epsilon,$$
$$Sx = U(x/g(x), (1-g(x))/\epsilon) \quad \text{if } 1 \geqslant g(x) \geqslant 1 - \epsilon.$$

(Here $g(x)$ is the Minkowski functional of $\mathscr{M}$; see 4.2.5.) By the lemmas below, for $\epsilon$ sufficiently small $S$ has no fixed points. However, $S$ maps $\partial\mathscr{M}$ into $\mathscr{M}$ since if $x \in \partial\mathscr{M}$, $g(x) = 1$ and $Sx = U(x, 0) = U_0 x \in \mathscr{M}$ by (*A*). Also $S$ is compact so that we have a contradiction to Rothe's theorem 4.2.4. Thus the assumption that $U_1$ had no fixed point must have been false. $\square$

**L E M M A** 1. *For $\epsilon$ sufficiently small, the equation*

$$U_1(x/(1-\epsilon)) = x$$

*has no solutions for*          $g(x) \leqslant 1 - \epsilon$.

*Proof.* The condition $g(x) \leqslant 1 - \epsilon$ means just that

$$x/(1-\epsilon) \in \mathscr{M}.$$

Suppose the lemma false, then we could find a real sequence $\epsilon_n \to 0$ and a sequence $x_n \in \mathcal{M}$ such that

$$U_1(x_n/(1-\epsilon_n)) = x_n. \qquad (1)$$

By the compactness of $U$ we can assume that $x_n$ converges, say $x_n \to y$; then by the continuity of $U_1$, (1) gives $U_1 y = y$, contradicting the assumption that $U_1$ has no fixed point. $\square$

LEMMA 2. *For $\epsilon$ sufficiently small, the equation*

$$U(x/g(x), (1-g(x))/\epsilon) = x$$

*has no solution for* $\qquad 1 \geqslant g(x) \geqslant 1 - \epsilon$.

*Proof.* Suppose on the contrary that we have sequences $\epsilon_n \to 0$ and $x_n$ such that $1 \geqslant g(x_n) \geqslant 1 - \epsilon_n$ and

$$U(x_n/g(x_n), \quad (1-g(x_n))/\epsilon_n) = x_n. \qquad (2)$$

Then we have $1 \geqslant (1-g(x_n))/\epsilon_n \geqslant 0$. We can assume without loss of generality that $(1-g(x_n))/\epsilon_n \to t \in [0,1]$ and also, by the compactness of $U$, that $x_n \to y \in \mathcal{M}$. Thus (2) gives, by continuity,

$$U(y, t) = y.$$

This contradicts $(B)$ since $g(y) = \lim g(x_n) = 1$; that is, $y \in \partial \mathcal{M}$. $\square$

## 4.4  Krasnoselskii's theorem

We will consider the sum of a compact mapping and a contraction mapping. This combination can occur in practice; in dealing with a perturbed differential operator we may find that the perturbation leads to a contraction mapping while inversion of the differential operator gives a compact mapping. See for instance Schauder (1932).

THEOREM 4.4.1 (*Krasnoselskii*) *Let $\mathcal{M}$ be a closed convex non-empty subset of a Banach space $\mathcal{S}$. Suppose that $A$ and $B$ map $\mathcal{M}$ into $\mathcal{S}$ and that*
   (i)  $Ax + By \in \mathcal{M}$  $(\forall x, y \in \mathcal{M})$,
   (ii)  *$A$ is compact and continuous.*
   (iii)  *$B$ is a contraction mapping.*
*Then there exists $y$ in $\mathcal{M}$ such that*

$$Ay + By = y.$$

We require

LEMMA 4.4.2    *If $B$ is a contraction mapping of a subset $\mathcal{X}$ of a normed space $\mathcal{S}$ into $\mathcal{S}$ then $I - B$ is a homeomorphism on $\mathcal{X}$ to $(I - B)\mathcal{X}$. If $(I - B)\mathcal{X}$ is precompact then $\mathcal{X}$ is precompact.*

*Proof.* Clearly, $I - B$ is continuous. Also

$$\|(I - B)x - (I - B)y\| \geqslant \|x - y\| - \|Bx - By\| \geqslant (1 - k)\|x - y\|$$

(where $0 < k < 1$), so that $(I - B)^{-1}$ is continuous. The same inequality shows that $x_1, \ldots, x_n$ is an $\epsilon$-net for $\mathcal{X}$ if

$$(I - B)x_1, \ldots, (I - B)x_n$$

is a $(1 - k)$ $\epsilon$-net for $(I - B)\mathcal{X}$. $\square$

*Proof of theorem 4.4.1.* For each $y$ in $\mathcal{M}$, the equation

$$z = Bz + Ay$$

has a unique solution $z$ in $\mathcal{M}$, since $z \to Bz + Ay$ defines a contraction mapping of $\mathcal{M}$ into $\mathcal{M}$. Thus $z = (I - B)^{-1}Ay$ is in $\mathcal{M}$. By the lemma, $(I - B)^{-1}A$ is continuous and compact on $\mathcal{M}$ into $\mathcal{M}$. By Schauder's theorem 4.1.3, $(I - B)^{-1}A$ has a fixed point in $\mathcal{M}$; this point $y$ is the one required. $\square$

REMARKS

(1) For generalisations of the theorem see for instance Reinermann (1971 $a$).

(2) Sadovskii (1967) defines a condensing operator as one which reduces the value of

$$\chi(\mathcal{M}) = \inf\{\epsilon > 0: \mathcal{M} \text{ has a finite } \epsilon\text{-net}\}$$

whenever $\mathcal{M}$ is bounded and not precompact. An interesting argument extends 4.4.1 to condensing operators (exercise 7).

## 4.5   Locally convex spaces

The basic result for these spaces is an extension of Schauder's theorem 4.1.3.

THEOREM 4.5.1    *Let $\mathcal{E}$ be a locally convex space, $\mathcal{M}$ a convex non-empty subset of $\mathcal{E}$. Let $T$ be a continuous mapping of $\mathcal{M}$ into a compact subset $\mathcal{K}$ of $\mathcal{M}$. Then $T$ has a fixed point.*

*Sketch of proof.* (Essentially this is Singbal's proof, which is given in an appendix to Bonsall (1962).) Adapt the proof of 4.1.1 replacing 4.1.3 by:

*For each neighbourhood $\mathscr{V}$ of 0 there is a finite set $\mathscr{X}$ in $\mathscr{K}$ and a continuous mapping $P$ of $\mathscr{K}$ into $\mathrm{co}(\mathscr{X})$ such that*

$$Px - x \in \mathscr{V} \quad (x \in \mathscr{K}).$$

The proof can be modelled on 4.1.1; for details see Nagumo (1951 b). If $\mathscr{M}$ is closed we can complete the proof by using Nagumo's theorem 1, which asserts that $\mathscr{R}(T - I)$ is closed. To obtain the general case, prove the following analogue of 1.1.3:

*Let $\mathscr{E}$ be a topological vector space. Suppose that $T$ is a continuous mapping of a subset $\mathscr{M}$ of $\mathscr{E}$ into a compact set $\mathscr{K} \subset \mathscr{M}$. Suppose that for each neighbourhood $\mathscr{V}$ of 0 there is a point $x_{\mathscr{V}}$ such that $Tx_{\mathscr{V}} - x_{\mathscr{V}} \in \mathscr{V}$. Then $T$ has a fixed point.* $\square$

Our results 4.3.2, 4.2.4 and 4.3.3 were originally given, by Schaefer (1955) and Potter (1973), for locally convex spaces.

## Exercises

1. We say that a set $\mathscr{M}$ has the *compact fixed point property* if each continuous mapping of $\mathscr{M}$ into a compact subset of $\mathscr{M}$ has a fixed point. Show that

(i) if a set $\mathscr{X}$ has this property then any retract of $\mathscr{X}$ has this property;

(ii) any convex set in a normed space has this property;

(iii) the unit sphere in an infinite-dimensional Hilbert space has this property (use exercise 2 of Chapter 2).

2. (i) With $\mathscr{K}$ and $\mathscr{X}$ as in 4.1.3 show that each point of $\mathrm{co}(\mathscr{K})$ lies within $\epsilon$ of some point of $\mathrm{co}(\mathscr{X})$.

(ii) If $\mathscr{K}$ is a compact subset of a Banach space show that $\overline{\mathrm{co}}(\mathscr{K})$ is compact.

(iii) Prove 4.1.1 in the case where $\mathscr{B}$ is complete and $\mathscr{M}$ is closed.

3. (*Altman*) Let $\mathscr{M}$ be the closed unit ball and $\mathscr{S}$ the unit sphere in a normed space $\mathscr{B}$. Let $F$ map $\mathscr{M}$ into a compact subset of $\mathscr{B}$ and assume that

$$\|x - F(x)\|^2 \geqslant \|F(x)\|^2 - \|x\|^2 \quad (x \in \mathscr{S}).$$

Show that $F$ has a fixed point in $\mathcal{M}$. (Use 4.3.3, with

$$U(x,t) = tF(x).)$$

4. If $\mathcal{K}$ is precompact rather than compact,

    (i) 4.1.3 remains true; hence

    (ii) in 4.1.1 we can conclude that $T$ has $\epsilon$-fixed points for all $\epsilon > 0$.

5. A compact operator or a contraction mapping is condensing; the sum of any two condensing operators is condensing.

6. $\chi(\mathcal{M}) = \chi(\overline{\mathcal{M}}) = \chi(\mathrm{co}\,\mathcal{M})$ for any bounded set $\mathcal{M}$.

7. Let $\mathcal{S}$ be a bounded closed convex subset of a Banach space and $T$ a condensing mapping of $\mathcal{S}$ into $\mathcal{S}$. Choose $x$ in $\mathcal{S}$ and show that

    (i) $\chi(\{x, Tx, T^2x, \ldots\}) = \chi(\{Tx, T^2x, T^3x, \ldots\})$.

    (ii) $\mathcal{Y} = \{x, Tx, T^2x, \ldots\}$ is precompact.

    (iii) If $\mathcal{K}$ is the set of limit points of $\mathcal{Y}$ then $\mathcal{K}$ is compact nonempty and $T\mathcal{K} = \mathcal{K}$.

    (iv) There is a minimal closed convex subset $\mathcal{L}$ of $\mathcal{S}$ containing $\mathcal{K}$ and invariant under $T$.

    (v) $\mathcal{L} = \overline{\mathrm{co}}(T\mathcal{L})$; $\chi(\mathcal{L}) = \chi(T\mathcal{L})$; $\mathcal{L}$ is compact.

    (vi) (*Sadovskii*). $T$ has a fixed point.

# 5. *Non-expansive mappings*

This chapter deals with various generalisations of the concept of a contraction mapping; the most important of these is the concept of a non-expansive mapping.

## 5.1 Bounded convex sets

DEFINITION 5.1.1  A map $T$ of a metric space into a metric space satisfying the condition

$$\rho(Tx, Ty) \leqslant \rho(x, y) \quad (\forall x, y)$$

is said to be *non-expansive*.

Contraction mappings, isometries and orthogonal projections are all non-expansive mappings. The study of non-expansive mappings has been one of the main features in recent developments of fixed point theory: see for instance Browder (1973). There are some interesting unsolved problems. We will sketch (§6.4) an application to non-linear equations of evolution.

A non-expansive mapping of a complete space need not have a fixed point (consider a translation operator $f \to f + g$ with $g \neq 0$, in a Banach space). A fixed point of a non-expansive mapping need not be unique (consider $T = I$).

THEOREM 5.1.2  *Let $\mathcal{M}$ be a bounded closed convex subset of a Banach space and $T$ a non-expansive mapping of $\mathcal{M}$ into $\mathcal{M}$. Then for each $\epsilon > 0$, $T$ has an $\epsilon$-fixed point $x(\epsilon)$ in $\mathcal{M}$; that is,*

$$\|Tx(\epsilon) - x(\epsilon)\| < \epsilon.$$

*Proof.* Assume without loss of generality that $0 \in \mathcal{M}$ and that $\mathcal{M}$ is contained in a ball of radius $R$ about 0. For each $r < 1$, consider the fixed point $y$ of the contraction mapping $rT$ of $\mathcal{M}$ into $\mathcal{M}$. We have

$$\|Ty - y\| = \|Ty - rTy\| = (1 - r)\|Ty\| \leqslant (1 - r)R < \epsilon,$$

for $r$ sufficiently close to 1. $\square$

3-2

If $\mathscr{M}$ is compact, the previous result and 1.1.3 yield a fixed point for $T$. Thus we have an elementary proof of Schauder's theorem 2.3.7 for the case of non-expansive mappings. In a similar way we can produce fixed points for non-expansive mappings of a compact star-shaped body, a compact cone, or any other compact set whose identity mapping can be uniformly approximated by contraction mappings.

It is remarkable that the following result does not assume compactness.

**THEOREM** 5.1.3 (*Browder*, 1965*b*) *Let $\mathscr{M}$ be a bounded closed convex subset of a Hilbert space $\mathscr{H}$. Then any non-expansive mapping $T$ of $\mathscr{M}$ into $\mathscr{M}$ has a fixed point.*

*Proof.* By 5.1.2, there exist points $x(\epsilon)$ such that

$$\|(I-T)x(\epsilon)\| < \epsilon \quad (\forall \epsilon > 0).$$

Thus if we can show that $(I-T)\mathscr{M}$ is closed, we will have $0 \in (I-T)\mathscr{M}$ and $0 = x - Tx$ for some $x \in \mathscr{M}$. The proof that $(I-T)\mathscr{M}$ is closed is given in 5.1.7 to 5.1.9 below. $\square$

**REMARK** 5.1.4 The result remains true for the case of a uniformly convex Banach space; see Browder (1965*d*), Kirk (1965) or Goebel (1969). The cases

(*a*) where $\mathscr{H}$ is any reflexive Banach space,

(*b*) where $\mathscr{H}$ is any strictly convex Banach space, or

(*c*) where $\mathscr{M}$ is a weakly compact and convex subset of any Banach space

all appear to be open.

By 2.4.3 we cannot allow a Lipschitz constant greater than 1.

**EXAMPLE** 5.1.5 (*Beals*) *If $\mathscr{M}$ is the unit ball of the sequence space $(c_0)$ there is a non-expansive fixed-point-free mapping of $\mathscr{M}$ into $\mathscr{M}$.*

*Proof.* Put $T(x_1 \; x_2 \; ...) = (1 \; x_1 \; x_2 \; ...)$. $\square$

In the following lemmas, needed for 5.1.3 we will assume (without loss of generality) that $\mathscr{H}$ is a real Hilbert space.

**DEFINITION** 5.1.6 A mapping $F$ of a Hilbert space $\mathscr{H}$ into $\mathscr{H}$ is *monotone* if $F$ is continuous and we have

$$(Fx - Fy, x - y) \geqslant 0 \quad (Ax, y \in \mathscr{H}).$$

For our application, $F$ will be continuous for the strong topology; however, a far weaker continuity property of $F$ is actually needed (in 5.1.8).

LEMMA 5.1.7   *Let $\mathcal{M}$ be a closed convex subset of a Hilbert space $\mathcal{H}$. If $T$ is a non-expansive mapping of $\mathcal{M}$ into $\mathcal{H}$ then $I - T$ is the restriction to $\mathcal{M}$ of a monotone operator.*

*Proof.* Let $r$ be the metric retraction of $\mathcal{H}$ onto $\mathcal{M}$; each point of $\mathcal{H}$ is mapped onto the nearest point of $\mathcal{M}$ (see 2.1.4). Since $r$ is non-expansive, $Tr$ is a non-expansive mapping of $\mathcal{H}$ into $\mathcal{H}$. For $x, y \in \mathcal{H}$,

$$((I - Tr)x - (I - Tr)y, x - y)$$
$$= \|x - y\|^2 - (Trx - Try, x - y)$$
$$\geqslant \|x - y\|^2 - \|Trx - Try\| \cdot \|x - y\| \geqslant 0,$$

so that $I - Tr$ is monotone. The restriction to $\mathcal{M}$ of $I - Tr$ is $I - T$. $\square$

LEMMA 5.1.8   *If $F$ is monotone and $u_0$ and $v_0$ are elements of $\mathcal{H}$ such that*
$$(Fu - v_0, u - u_0) \geqslant 0 \quad (\forall u \in \mathcal{H}) \tag{*}$$
*then $v_0 = Fu_0$.*

*Proof.* For any $v$ in $\mathcal{H}$ and any $t > 0$ write $u_t = u_0 + tv$. From (*) with $u = u_t$ we obtain $(Fu_t - v_0, v) \geqslant 0$ so that

$$(Fu_t - Fu_0, v) \geqslant (v_0 - Fu_0, v).$$

Let $t \to 0 +$; then $Fu_t \to Fu_0$ so that we obtain

$$0 \geqslant (v_0 - Fu_0, v) \quad (\forall v \in \mathcal{H}).$$

Clearly $v_0 = Fu_0$. $\square$

LEMMA 5.1.9   *If $F$ is monotone on $\mathcal{H}$ and $\mathcal{M}$ a bounded closed convex subset of $\mathcal{H}$ then $F\mathcal{M}$ is closed.*

*Proof.* Suppose that $u_n \in \mathcal{M}$ for $n = 1, 2, \ldots$ and that $Fu_n \to v_0$. We may assume without loss of generality that $u_n$ converges weakly to $u_0 \in \mathcal{M}$. For all $u \in \mathcal{H}$,

$$(Fu - Fu_n, u - u_n) \geqslant 0.$$

Letting $n \to \infty$,        $(Fu - v_0, u - u_0) \geqslant 0,$

so that $v_0 = Fu_0$ by 5.1.8. $\square$

For further results on non-expansive mappings (including an analogue of Rothe's theorem 4.2.3, and a continuation theorem) see Browder (1965 $b$). For generalisations to Banach spaces see for instance Kirk (1970) or Browder (1973); the latter reference also discusses generalisations of the concept of a monotone operator.

## 5.2  Various

The heterogeneous ideas considered in this section all have some relation to the idea of a contraction mapping. Some of the mappings considered are non-expansive. Our first result generalises Banach's theorem 1.2.2 to the case where some iterate of $T$ is a contraction mapping (this occurs, for example, if $T$ is a Volterra integral operator).

THEOREM 5.2.1   *Let $\mathcal{M}$ be a complete non-empty metric space and $T$ a mapping of $\mathcal{M}$ into $\mathcal{M}$ such that $T^K$ is a contraction mapping (for some integer $K > 1$). Then $T$ has a unique fixed point in $\mathcal{M}$.*

*Proof.* Let $z$ be the unique fixed point for $T^K$. Then

$$T^K(Tz) = T(T^Kz) = Tz$$

so that $Tz$ is also a fixed point. By uniqueness $Tz = z$. $\square$

DEFINITION 5.2.2   We will call $T$ a *shrinking mapping* if

$$\rho(Tx, Ty) < \rho(x, y) \quad (x \neq y).$$

(The term 'contractive mapping' has been used for the same concept.)

Thus a shrinking mapping is non-expansive, but need not be a contraction mapping. Clearly, a shrinking mapping can have at most one fixed point.

THEOREM 5.2.3   *Any shrinking mapping $T$ of a compact non-empty metric space $\mathcal{M}$ into itself has a fixed point.*

*Proof.* Since $\rho(Tx, x)$ is continuous and $\mathcal{M}$ is compact there exists $z$ in $\mathcal{M}$ such that

$$\rho(Tz, z) = \inf_{x \in \mathcal{M}} \rho(Tx, x). \tag{1}$$

Thus $Tz = z$ since otherwise we would have

$$\rho(T^2z, Tz) < \rho(Tz, z)$$

contradicting (1). □

Some other results on shrinking mappings are given in Edelstein (1962). The following question appears to be open.

PROBLEM   Does every shrinking mapping of the closed unit ball in a Banach space have a fixed point?

We now give a 'local' version of the contraction mapping theorem, following Edelstein (1961) (see also Bailey (1966)).

DEFINITION 5.2.4   A mapping $T$ is *locally* $(\epsilon, \lambda)$ *a contraction mapping* if:

We have            $\epsilon > 0$   and   $0 < \lambda < 1$          $\Big\}$   (2)
and        $\rho(Tx, Ty) \leqslant \lambda\rho(x, y)$   wherever   $\rho(x, y) < \epsilon.$

Edelstein gives the following example showing that such a mapping need not be a contraction mapping:

$$\mathcal{M} = \{e^{i\theta} : 0 \leqslant \theta \leqslant \tfrac{3}{2}\pi\}, \quad Te^{i\theta} = e^{i\theta/2}.$$

THEOREM 5.2.5 (*Edelstein*)   *Let $\mathcal{M}$ be a complete $\epsilon$-chainable metric space and $T$ a mapping of $\mathcal{M}$ into $\mathcal{M}$ satisfying (2). Then $T$ has a unique fixed point in $\mathcal{M}$.*

*Sketch of proof.* Choose $x$ in $\mathcal{M}$. Take points

$$x = x_0, x_1, x_2, \ldots, x_m = Tx$$

in $\mathcal{M}$ such that $\rho(x_i, x_{i+1}) < \epsilon$. From (2),

$$\rho(T^r x_i, T^r x_{i+1}) \leqslant \lambda^r \rho(x_i, x_{i+1}) < \lambda^r \epsilon < \epsilon/m$$

for an integer $r$ sufficiently large. Thus $\rho(T^r x, T^{r+1} x) < \epsilon$. It can now be shown (as in 1.2.2, with $y = T^r x$) that $T^{n+r} x$ converges to a fixed point as $n \to \infty$. □

Edelstein gives an application of 5.2.5 to a problem on analytic functions.

Converses of the contraction mapping theorem have been discussed by Meyers (1967), Janos (1967) and Edelstein (1969).

If $\cap\, T^n \mathcal{M}$ is a one-point set, and other conditions are satisfied, then $\mathcal{M}$ can be (re-)metrised in such a way that $T$ is a contraction mapping. For several commuting mappings in $\mathcal{M}$, see Meyers (1970).

In recent years many papers by various authors have generalised the concepts of contraction mapping and non-expansive mapping to topological spaces.

### Exercises

1. Let $\mathcal{M}$ be a convex subset of a normed space $\mathscr{S}$. Let $R$ be a non-expansive mapping on $\mathcal{M}$ into $\mathscr{S}$. Then for $0 < t < 1$ the mapping

$$S_t = tI + (1-t)R$$

is non-expansive and has the same set of fixed points as $R$. If $R\mathcal{M} \subset \mathcal{M}$ then $S_t\mathcal{M} \subset \mathcal{M}$. (See Browder and Petryshyn (1966) for references concerning $S_t$.)

2. If $\mathcal{M}$ is a closed convex subset of a Hilbert space $\mathscr{S}$, prove the following statements.

(i) The radial retraction of $\mathscr{S}$ onto $\mathcal{M}$ (see 4.2.6) need not be non-expansive.

(ii) The metric retraction of $\mathscr{S}$ onto $\mathcal{M}$ (see 2.1.4) is non-expansive.

(iii) If $T$ is a contraction mapping of $\mathcal{M}$ into $\mathscr{S}$ such that $T(\partial\mathcal{M}) \subset \mathcal{M}$ then $T$ has a fixed point. (See Vidossich (1971) for a partial extension to Banach spaces.)

(iv) In (iii) we can replace 'contraction' by 'non-expansive' if $\mathcal{M}$ is a closed ball.

3. In 5.1.2, we obtain $\epsilon$-fixed points even if $\mathcal{M}$ is not closed or the space is not complete.

# 6. Existence theorems for differential equations

We intend to give a sketch of some methods by which existence theorems have been obtained from fixed point theorems, and an introduction to the literature. As general references we mention Miranda (1955) and Cronin (1964) for earlier work and Browder (1973) for recent work.

## 6.1 Methods available

Throughout this chapter, 'solving' means proving that a solution exists. Fixed point methods are most important in solving non-linear problems. There are several ways to reduce a non-linear existence problem to a fixed point problem (for a mapping in function space). We give first the most useful method.

METHOD 6.1.1. *The linearisation trick* of Leray and Schauder (1934). Suppose that an expression $D(f, g)$ (which could involve $f$ and $g$ in any way) is linear in $f$. (In the cases that interest us, $D(f, g)$ will involve derivatives of $f$ and possibly of $g$.) Suppose also that the linear equation

$$D(f, g) = 0 \tag{1}$$

has a unique solution $f = Tg$ for each $g$ in some set $\mathcal{M}$. Then to find a solution in $\mathcal{M}$ of the (usually non-linear) equation

$$D(f, f) = 0 \tag{2}$$

is equivalent to finding a fixed point (in $\mathcal{M}$) of the mapping $T$. Thus a particular non-linear equation can be studied by means of a more general linear equation, together with a fixed point problem.

Boundary conditions and differentiability conditions can be incorporated in the definition of the set $\mathcal{M}$. Thus we must know that (1) has a unique solution, subject to these conditions, and we deduce that (2) has a solution, subject to the same conditions.

[ 41 ]

For the method to work $T$ must be 'small' (i.e. usually a compact mapping or a contraction mapping). This appears to mean that (1) and (2) must involve the same derivatives of highest order; thus (1) must be rather mildly non-linear.

For example, to solve the non-linear equation

$$f'(t) + f(t)^2 = \sin t \qquad (2)'$$

subject to the condition $f(0) = 0$, we would take $\mathcal{M}$ as some set of functions satisfying $f(0) = 0$, and use either of the following linear equations

$$f'(t) + g(t)^2 = \sin t, \qquad (1)'$$

or

$$f'(t) + g(t)f(t) = \sin t. \qquad (1)''$$

METHOD 6.1.2 *Replace the D.E. and its boundary conditions by an integral equation*

$$y = J(y). \qquad (3)$$

The solution is then given by a fixed point of the mapping $y \to J(y)$. The operator $J$ usually involves a 'Green's function' obtained by some special trick. More systematically an equation of the type (3) emerges if the operator $T$ mentioned in method 6.1.1 can be expressed as an integral operator (see §§ 6.2, 6.3, 6.5). It may however be harder to obtain the explicit expression for the mapping as an integral operator $J$ than to apply method 6.1.1 directly (see §6.3). It may be useful to have the integral formula for $J$ if we wish to show that $J$ gives a contraction mapping.

In most cases it will not be easy to reduce a D.E. to the form (3).

METHOD 6.1.3 *The continuation methods* of §4.3 are often used in conjunction with the linearisation trick; see §6.7. In this approach we need not find a set $\mathcal{M}$ which is mapped into itself by the mapping $U_1$ which really interests us. (The set $\mathcal{M}$ is usually mapped into itself by the simpler mapping $U_0$.)

METHOD 6.1.4 *Periodic solutions* of equations with periodic data can sometimes be found by a method given in §6.4.

## 6.2   Ordinary D.E.s

We write $y'$ for $dy/dt$. Consider a system of ordinary D.E.s which can be written as a first-order system in the form:

$$\left.\begin{aligned}
y_1' &= f_1(t, y_1, y_2, \ldots, y_n), \\
&\;\vdots \\
y_n' &= f_n(t, y_1, y_2, \ldots, y_n),
\end{aligned}\right\} \tag{1}$$

or more briefly $\qquad Y'(t) = F(t, Y(t)), \tag{2}$

where $\qquad Y = (y_1, \ldots, y_n) \quad \text{and} \quad F = (f_1, \ldots, f_n).$

We will use any of the usual norms on $R^n$ and write $|X|$ for the norm of $X$.

We will look for a solution of (2) such that

$$Y(a) = B, \tag{3}$$

where the real number $a$ and the point $B$ in $R^n$ are given. We will assume that $F$ is continuous in $Y$ and $t$, i.e. that each $f_r$ is jointly continuous in $y_1, \ldots, y_n$ and $t$.

Use the linearisation trick: to find a solution of (2), (3) in some set $\mathcal{M}$ we consider, for each $X$ in $\mathcal{M}$, the solution $Y = UX$ of the linear equation
$$Y' = F(t, X) \tag{4}$$
subject to (3).

In all cases that interest us, $Y = UX$ is uniquely defined by (3), (4) and we have

$$UX(t) = B + \int_a^t F(s, X(s))\, ds. \tag{5}$$

Thus for a solution of (3), (2) we require a fixed point of the mapping $U$ given by (5).

THEOREM 6.2.1 (*Cauchy–Lipschitz*)   *Let $(a, B)$ be a point in an open subset $S$ of $R^1 \times R^n$. Suppose that*
  (i)   *the function $F(t, X)$ is continuous in $S$,*
  (ii)  *$F$ satisfies a Lipschitz condition of the form*

$$|F(t, X) - F(t, Y)| \leqslant K|X - Y|,$$

*for some real $K$ and all points $(t, X)$ and $(t, Y)$ in $S$. Then a unique solution of (2), (3) exists in some neighbourhood of $a$.*

*Proof.* We define a set $\mathcal{M}$ as in §1.3; however, the members of $\mathcal{M}$ are now functions from $R^1$ to $R^n$ with graphs in the subset $R = \overline{N}(a, d) \times \overline{N}(B, Ld)$ of $R^1 \times R^n$. As before we see that the mapping $U$ defined by (5) is a contraction mapping of $\mathcal{M}$ into $\mathcal{M}$, and has a unique fixed point. This gives a solution of (2), (3). We can check that any solution of (2) and (3) must in fact be in $\mathcal{M}$; thus the solution is unique. □

REMARK   The condition (ii) is equivalent to the statement that each function $f_i$ satisfies a Lipschitz condition with respect to each variable $y_j$.

REMARK   To obtain a global result from 6.2.1, we must make successive extensions of the solution; Cronin (1964, theorem II.1.7) shows that the solution can be extended to the boundary of $S$. The method of Bielecki (1956), given in Edwards (1965), shows that where $S$ is an infinite strip $(a - \delta, a + \delta) \times (-\infty, \infty)$ we can find a unique solution of (2), (3), valid throughout $(a - \delta, a + \delta)$ by using the norm

$$N(X) = \sup \{e^{-cK|t-a|} |X(t)| : t \in (a - \delta, a + \delta)\},$$

where $c$ is any fixed number greater than 1. In terms of this norm, (5) defines a contraction mapping $U$ in $C(a - \delta, a + \delta)$. This 'global' method is only applicable where the Lipschitz condition holds in an infinite strip and is not applicable to an equation such as $y' = y^2$. Stokes (1960) obtains solutions of (2), (3) for $0 < t < \infty$; the solutions belong to various spaces of functions on $(0, \infty)$.

THEOREM 6.2.2 (*Peano*)   *If $F(t, Y)$ is a continuous function of $t$ and $Y$ in a neighbourhood of $(a, B)$ then the problem (2), (3), i.e.*

$$Y' = F(t, Y), \quad Y(a) = B,$$

*has at least one solution in a neighbourhood of $a$.*

*Proof.* We can assume that $|F(t, Y)| \leqslant K$ for $|t - a| \leqslant \epsilon$ and $|Y - B| \leqslant \epsilon$. Choose $\delta$ so that $\delta \leqslant \epsilon$ and $\delta K \leqslant \epsilon$. Let $\mathcal{S}$ be the space of continuous functions defined for $|t - a| \leqslant \delta$, with values in $R^n$; we use the uniform norm. Let $\mathcal{M}$ be the closed convex subset of $\mathcal{S}$ consisting of functions $Y$ with $|Y(t) - B| \leqslant \delta K$ for all $t$. Define $U$ by (5); then $U$ maps $\mathcal{M}$ into $\mathcal{M}$ since for $Y$ in $\mathcal{M}$ and $|t-a| \leqslant \delta$,

$$|UY(t) - B| = \left| \int_a^t F(t, Y(t)) \, dt \right| \leqslant \delta K.$$

$U$ is continuous since for $Y, Z \in \mathcal{M}$ and $\|Y - Z\| \to 0$,

$$\|UY - UZ\| = \sup_t \left| \int_a^t F(s, Y(s)) - F(s, Z(s)) \, ds \right| \to 0,$$

by the uniform continuity of $F$. Also

$$\|Uy\| \leqslant |B| + \delta K$$

and $$|UY(s) - UY(t)| \leqslant \left| \int_s^t F(t, Y(t)) \, dt \right| \leqslant K|s - t|.$$

Thus $U\mathcal{M}$ is a uniformly bounded and equicontinuous family of functions, so that $U\mathcal{M}$ is precompact. By Schauder's theorem 4.1.1, $U$ has a fixed point $Y$. This gives the solution of (2), (3). $\square$

Questions of analyticity can be treated as follows.

THEOREM 6.2.3 (*Picard*)   *If $f(t, y)$ is an analytic function of $t$ and $y$ in a neighbourhood of $(a, b)$ then the problem*

$$y' = f(t, y), \quad y(a) = b \tag{6}$$

*has a unique analytic solution in a neighbourhood of $a$.*

   *Proof.* We can assume that for $|t - a| \leqslant \epsilon$ and $|y - b| \leqslant \epsilon$ we have $|f| \leqslant K$ and $|\partial f/\partial y| \leqslant L$. Choose

$$\delta \leqslant \min(\epsilon, \epsilon/K, 1/(2L)).$$

Let $\mathcal{M}$ be the set of functions $y$ analytic for $|t - a| < \delta$, continuous for $|t - a| \leqslant \delta$ and such that $|y - b| \leqslant K\delta$ for $|t - a| \leqslant \delta$. Clearly $\mathcal{M}$ is complete in the uniform norm. Define the mapping $U$ by (5), interpreting the integral as a complex contour integral. Then $U$ maps $\mathcal{M}$ into $\mathcal{M}$. Finally, for $h$ and $g$ in $\mathcal{M}$,

$$\|Uh - Ug\| = \sup_z \left| \int_a^z f(t, h) - f(t, g) \, dt \right|$$
$$\leqslant \sup_z |z - a| L \sup_t |h(t) - g(t)|$$
$$\leqslant \delta L \|h - g\|$$
$$\leqslant \tfrac{1}{2} \|h - g\|.$$

Thus the contraction mapping theorem shows that $U$ has a unique fixed point in $\mathcal{M}$. This is the only solution of (6): any solution $y$ has a complex derivative in a neighbourhood of $a$, thus $y$ is analytic near $a$, thus $y$ is in $\mathcal{M}$ if we take $\delta$ small enough. $\square$

If we have a problem

$$y' = f(t, y, \lambda), \quad y(a) = b$$

we can consider a space of functions $y(t, \lambda)$ and obtain a solution which depends on the parameter $\lambda$ – analytically, if $f$ is analytic, or continuously if $f$ is continuous and satisfies a Lipschitz condition with respect to $y$.

## 6.3 Two-point boundary conditions

A problem which is discussed by Edwards (1965), following Bass (1958), is

$$\frac{d^2x}{dt^2} = f\left(t, x, \frac{dx}{dt}\right) \quad \text{for} \quad 0 \leqslant t \leqslant T, \tag{1}$$

with $x(0) = a$ and $x(T) = b$ given. The assumption is made that $f$ is continuous and bounded on $[0, T] \times R \times R$. The procedure is to reduce the problem to an integral equation

$$x(t) = a + \frac{(b-a)t}{T} - \int_0^T G(t, s) f(s, x(s), x'(s)) \, ds. \tag{2}$$

(The Green's function $G$ is given by a simple explicit formula.) Consider the mapping $U$ on $C^{(1)}$ to $C^{(1)}$ which maps each $x$ onto the right-hand side of (2). We require a fixed point for $U$. From the boundedness of $f$ and the formula for $Ux$ it follows that the set of points $Ux$ is precompact. Thus a fixed point exists by Schauder's theorem (4.1.4). For the details see the sources mentioned.

We now show that by using the linearisation trick, we can solve the problem without considering the Green's function. We define the mapping $U$ of $C^{(1)}$ into $C^{(1)}$ as follows: for $y$ in $C^{(1)}$, $Uy = x$ is the unique solution of

$$x''(t) = f(t, y(t), y'(t)), \quad x(0) = a, \quad x(T) = b$$

($x$ is found by two integrations, choosing the constants correctly.) To apply 4.1.4, we will show that $UC^{(1)}$ is precompact in $C^{(1)}$. If $x = Uy$ we have $|x''(t)| \leqslant K(\forall t)$. Since also $x(0) = a$, $x(T) = b$, a simple *ad hoc* argument gives bounds independent of $x$ for $|x'(t)|$ and $|x(t)|$. Having such bounds for $x, x'$ and $x''$, we see that the set of points $x = Uy$ is precompact in $C^{(1)}$. An elementary argument shows that $U$ is closed from $C^{(1)}$ to $C^{(1)}$. Thus $U$ is continuous, by the lemma below, and Schauder's theorem (4.1.4) gives a fixed point for $U$. This fixed point for $U$ is the required solution of (1).

LEMMA 6.3.1 (*'Closed Graph'*)   *Any closed single-valued mapping $U$ of a metric space $\mathcal{M}$ into a compact metric space $\mathcal{N}$ is continuous.*

*Proof.* If $x_n \in \mathcal{M}$ and $x_n \to y \in \mathcal{M}$, but $Ux_n \nrightarrow Uy$, we could assume without loss of generality that $Ux_n \to z \in \mathcal{N}$ with $z \neq Uy$. Thus we would have $\langle y, z \rangle$ in the graph of $U$, a contradiction. $\square$

## 6.4   Existence of periodic solutions

A fixed point method for finding periodic solutions of dynamical problems was used by Poincaré (1912); for proofs and further discussion see Birkhoff (1913, 1927). Essentially the same method is used in the following result.

THEOREM 6.4.1   *Consider the equation*

$$X'(t) = F(X, t) \tag{1}$$

*(where $X$ and $F$ take values in $R^n$). Assume that*

(i) *for each point $P_0$ in the closed ball $B^n$ there is a unique solution on $[0, \infty]$ such that $X(0) = P_0$;*

(ii) *if $X(0) \in B^n$ we have $X(t) \in B^n$ for all $t > 0$;*

(iii) *$X(t)$ depends continuously on $X(0)$;*

(iv) *$F$ has period $T > 0$ as a function of $t$.*

*Then the equation* (1) *has a solution of period $T$.*

*Proof.* For each $P_0$ in $B^n$ consider the point $P_T = X(T)$, for the solution $X(t)$ such that $X(0) = P_0$. By (ii), (iii) and Brouwer's theorem 2.1.11, the mapping $P_0 \to P_T$ has a fixed point, say $Z$. We consider the solution with $P_0 = P_T = Z$. Clearly this trajectory has period $T$ – for a proof see Cronin (1964, II.3.2). $\square$

REMARK For an application of this result to the question of periodic solutions of certain second-order (Sturm–Liouville) equations, see Cronin (1964, II.9.16).

A similar technique has been used by Browder (1965c, 1973) to obtain periodic solutions of the equation of evolution

$$\frac{du}{dt} + A(t)u = f(t, u). \tag{2}$$

Here each $A(t)$ is a linear operator in Hilbert space and $f(t, u)$ maps (a subset of) $R^1 \times \mathscr{H}$ into $\mathscr{H}$. Assumptions are made which ensure that (2) has a unique solution on $[s, \infty]$ for each $s \geqslant 0$ and each given value of $u(s)$. Further assumptions (of monotonicity) on the linear operators $A(t)$ and the non-linear operators $f(t, \cdot)$ ensure that the mapping $u(s) \to u(t)$ is non-expansive for $t > s$.

To obtain a periodic solution in the case where $A(t)$ and $f(t, u)$ have period $p$ as functions of $t$, the final assumption is made that for some $R > 0$,

$$\mathrm{Re}\,(f(t, u), u) < 0 \quad \text{for} \quad \|u\| = R, t \in [0, p].$$

This assumption is used in calculating $d\|u(t)\|^2/dt$, which turns out to be negative for $\|u(t)\| = R$. Thus a solution $u(t)$ which has entered the ball of radius $R$ must remain within that ball. Thus $u(0) \to u(p)$ gives a mapping of the ball into itself which has a fixed point by 5.1.3. The solution with $u(0) = u(p)$ is a solution of period $p$.

Güssefeldt (1970) obtains periodic solutions of the equation

$$x'(t) = F(x, t) \tag{3}$$

by a completely different approach: (3) is replaced by an integral equation

$$x(t) = \int_0^w G^A(t, s) f(x(s), s)\, ds \tag{4}$$

all solutions of which are periodic. Ordinary fixed point methods can be used to study (4): no attention need be paid to periodicity. The solutions obtained are the required periodic solutions of (3).

Periodic solutions of differential equations have also been derived from 'asymptotic' fixed point theorems; for references see Jones (1965).

## 6.5   Partial D.E.s: use of a Green's function

We consider a problem discussed by Nemyckii (1936). For other examples see Caccioppoli (1930).

We consider a bounded region $G$ of the plane with smooth boundary $\Gamma$, and closure $\bar{G} = G \cup \Gamma$. We write $P$ and $Q$ for points of $G$. We wish to solve the problem

$$\left.\begin{array}{c} \nabla^2 f(P) = -h(P, f(P)) \text{ in } G, \\ f = f_0 \text{ on } \Gamma, \end{array}\right\} \tag{1}$$

(where $\nabla^2$ is the Laplace operator $\partial^2/\partial x^2 + \partial^2/\partial y^2$, and $f_0$ is a given continuous function on $\Gamma$).

The discussion seems clearest if we use the linearisation trick. Consider, for any $g \in C(\bar{G})$, the linear equation

$$\left.\begin{array}{c} \nabla^2 f(P) = -h(P, g(P)) \text{ in } G, \\ f = f_0 \text{ on } \Gamma. \end{array}\right\} \tag{2}$$

LEMMA 6.5.1   *For each $g$ continuous in $\bar{G}$, each $f_0$ continuous on $\Gamma$, there exists a unique solution $f$ of the equation (2), which is given explicitly by*

$$f(P) = \phi(P) + \iint_G K(P, Q) h(Q, g(Q)) \, dQ, \tag{3}$$

*where $\phi$ is the solution of the problem*

$$\nabla^2 \phi = 0 \text{ in } G, \quad \phi = f_0 \text{ on } \Gamma \tag{4}$$

*and $K$ is the 'Green's function' of the region $G$.*

*Sketch of proof.* Since (2) is linear, the solution can be found as the sum of the solution $\phi$ of (4), and the solution of

$$\nabla^2 f = -h(P, g(P)) \text{ in } G, \quad f = 0 \text{ on } \Gamma. \tag{5}$$

The solution of (5) is given in terms of the Green's function $K$ by

$$f = \iint_G -K(P, Q) h(Q, g(Q)) \, dQ.$$

Thus the solution of (2) has the form (3). □

Consider the complete metric space $\mathcal{M} = C(\bar{G})$. For $g \in \mathcal{M}$, let $f = Ug$ be the solution of (2). The lemma ensures that the mapping $U$ is well defined and that $f = Ug$ is given explicitly by the formula (3). Using (3) we see that $U$ maps $\mathscr{M}$ into $\mathscr{M}$ and that for $g$ and $j$ in $\mathscr{M}$ we have

$$\|Ug - Uj\| = \sup_P \left| \iint_G K(P,Q)\,(h(Q,g(Q)) - h(Q,j(Q)))\,dQ \right|$$

$$\leqslant \sup_P \left| \iint_G K(P,Q)\,dQ \right| \sup_Q |h(Q,g(Q)) - h(Q,j(Q))|.$$

$U$ is thus a contraction mapping if $h$ satisfies a Lipschitz condition

$$|h(Q,f) - h(Q,g)| \leqslant \theta |f - g|,$$

where
$$\theta < \left( \sup_P \left| \iint_G K(P,Q)\,dQ \right| \right)^{-1}, \tag{6}$$

i.e. if the function $h$ is not greatly affected by $f$.

We thus obtain

Theorem 6.5.2 *If condition* (6) *is satisfied, the problem* (1) *has a unique solution.*

*Proof.* We have seen that given (6) the mapping $U$ is a contraction mapping, so has a unique fixed point; this point is the required solution. $\square$

## 6.6　The linearisation trick for partial D.E.s

We consider a plane region $G$ with boundary $\Gamma$. The linearisation trick was developed by Leray and Schauder for partial D.E.s. Consider for instance the equation

$$a(x,y,z)z_{xx} + b(x,y,z)z_{xy} + c(x,y,z)z_{yy} = 0 \quad \text{in } G, \tag{1}$$

with boundary conditions $\phi(z) = 0$ on $\Gamma$.

We consider for an arbitrary function $w$ the solution $z = Uw$ of the linear equation

$$a(x,y,w)z_{xx} + b(x,y,w)z_{xy} + c(x,y,w)z_{yy} = 0$$

subject to $\phi(z) = 0$ on $\Gamma$.

Conditions on $G$, $\Gamma$, $a$, $b$, $c$ (of continuity, ellipticity, etc.) are chosen to ensure that $U$ is well defined and precompact. The difficulty of finding a set $\mathscr{M}$ of functions which is transformed into itself led Leray and Schauder (1934) to use a continuation argument. Later writers, in particular Nirenberg (1953), were able to give a closed convex set $\mathcal{M}$ which is transformed into itself by $U$. Thus Schauder's theorem (4.1.1) gives a fixed point of $U$, in other words a solution for (1). For elliptic equations of the form (1), Nirenberg's method is presented in Courant and Hilbert (1962, IV.9). Nirenberg deals with the more general elliptic case where the coefficients $a$, $b$ and $c$ depend on $x$, $y$, $z$, $z_x$ and $z_y$. See Cronin (1964) for further discussion and references to other applications.

## 6.7 The methods of Leray–Schauder and Schaefer

The theorem of Leray and Schauder (discussed later; see 10.3.10) is usually applied according to the following scheme: for $0 \leqslant \lambda \leqslant 1$ we consider a (non-linear) differential equation; for instance we want $z$ to satisfy

$$a(x, y, z, \lambda) z_{xx} + b(x, y, z, \lambda) z_{xy} + c(x, y, z, \lambda) z_{yy} = 0 \qquad (1)$$

in a region $G$ of the $(x, y)$-plane; and

$$z = \phi(x, y, \lambda) \qquad (2)$$

on the boundary $\Gamma$. (The coefficients $a$, $b$, $c$ could also involve $z_x$ and $z_y$; this makes no essential difference.) The method is to replace (1) by the linear equation

$$a(x, y, w, \lambda) z_{xx} + b(x, y, w, \lambda) z_{xy} + c(x, y, w, \lambda) z_{yy} = 0 \qquad (3)$$

(if $z_x$ and $z_y$ occur in the coefficients $a$, $b$, $c$, we replace them by $w_x$ and $w_y$). The conditions placed on $a$, $b$, $c$, $y$ (such as continuity, ellipticity) ensure that for each $w$ in some set $\mathcal{M}$ of functions, and each $\lambda$, (2) and (3) have a unique solution $z$ (but this solution might not be in $\mathcal{M}$).

We consider the mapping $U_\lambda: w \rightarrow z$. A fixed point for $U_\lambda$ is a solution of (1) and (2). The problem we want to solve is the problem with $\lambda = 1$; we arrange the dependence on $\lambda$ to ensure that $\lambda = 0$ gives an easy problem. The Leray–Schauder theorem is then used to show that the problem with $\lambda = 1$ also has a solution. This method was first used by Leray and Schauder (1934); for other references see Leray (1950) and Cronin (1964).

Schaefer observed (1955) that the Leray–Schauder method was usually applied in the following way: the coefficient functions $a$, $b$, $c$ are taken independent of $\lambda$, being the coefficients of the problem we actually wish to solve; while the boundary value condition is taken as

$$z = \lambda\phi(x, y)$$

(where $\phi(x, y)$ is the boundary value function of the problem we wish to solve). (The advantage of this approach is that the problem with $\lambda = 0$ is very easy to discuss.) Schaefer pointed out that in this case we have $U_\lambda = \lambda T$ (where $T = U_1$) so that Schaefer's theorem 4.3.2 can be used in place of the Leray–Schauder theorem in these applications. See Cronin (1964, IV.9) for cases where this simplification could be used.

# 7. *Fixed points for families of mappings*

## 7.1 Commuting mappings

Consider a family $\mathscr{A}$ of mappings $T$ of some set into itself. If $Tx = x$ for all $T$ in $\mathscr{A}$ we say that $x$ is a common fixed point for $\mathscr{A}$ or for the mappings $T$ in $\mathscr{A}$. In this chapter we are concerned with the existence of common fixed points for families of mappings.

DEFINITION 7.1.1 We write $F(T)$ for the set of fixed points of $T$.

Clearly, the set of common fixed points of a family $\mathscr{A}$ is given by

$$\bigcap_{T \in \mathscr{A}} F(T).$$

A basic fact, for the study of common fixed points, is the following.

THEOREM 7.1.2 *If $T$ and $S$ map a set $\mathscr{M}$ into itself and $ST = TS$ then $SF(T) \subset F(T)$ and $S(T\mathscr{M}) \subset T\mathscr{M}$.*
*Proof.* If $Tx = x$, then $T(Sx) = S(Tx) = Sx$. If $z \in T\mathscr{M}$, $z = Tx$ then $Sz = STx = T(Sx) \in T\mathscr{M}$. $\square$

COROLLARY 7.1.3 *If $S$ and $T$ map a set $\mathscr{M}$ into $\mathscr{M}$ and $ST = TS$, and if $T$ has a unique fixed point $y$ then $y$ is also a fixed point for $S$.*
*Proof.* By 7.1.2, $Sy$ is a fixed point for $T$; by the uniqueness assumption, $Sy = y$. $\square$

For many years it was unknown whether two commuting (non-linear) mappings of a compact convex set into itself necessarily had a common fixed point. In view of 7.1.9, DeMarr (1963b) and other results, counter-examples are hard to construct; however, two workers independently solved the problem – using a computer in one case !

THEOREM 7.1.4 *There exist two commuting continuous mappings of* $[0, 1]$ *into itself without a common fixed point.*

*Proof.* See Boyce (1969) or Huneke (1969). □

In view of 7.1.4, it is not surprising that most of the results of this chapter involve a restriction to affine mappings. (A mapping $T$ of a convex set $\mathscr{M}$ is affine if it satisfies the identity

$$T(kx + (1-k)y) = kTx + (1-k)Ty$$

whenever $0 < k < 1$, $x \in \mathscr{M}$ and $y \in \mathscr{M}$.) We remark that, for affine mappings, fixed points have a natural geometric significance, since we can make an affine mapping linear by choosing a fixed point for a new origin. Thus a common fixed point for a family of affine mappings means a possible origin with respect to which all the mappings are linear.

THEOREM 7.1.5 (*Markov–Kakutani*) *Let $\mathscr{M}$ be a compact convex non-empty subset of a locally convex space. Let $\mathscr{A}$ be a commuting family of affine continuous mappings of $\mathscr{M}$ into $\mathscr{M}$. Then there exists a common fixed point for the mappings in $\mathscr{A}$.*

The original proof (due to Markov (1936)) depended on Tychonoff's theorem 2.3.8, i.e. on Brouwer's theorem 2.1.11. However, since we are only concerned with affine mappings, the present theorem is not as deep as Brouwer's theorem. A direct proof of 7.1.5 was given by Kakutani (1938). We give versions of both proofs.

We remark that Edwards (1965) gives a result which includes both 7.1.5 and the case of a soluble group of affine mappings. In §8.4 we discuss Day's theorem which is still more general.

LEMMA 7.1.6 $F(T)$ *is compact convex non-empty for each* $T \in \mathscr{A}$.

*Proof.* It is easy to check that $F(T)$ is convex and closed. Hence $F(T)$ is compact. By Tychonoff's theorem 2.3.8, $F(T)$ is non-empty. □

LEMMA 7.1.7 $F(T) \cap F(S)$ *is compact convex non-empty for all* $S$, $T$ *in* $\mathscr{A}$.

*Proof.* By 7.1.2, $S$ maps $F(T)$ into itself. Clearly, the required set $F(S) \cap F(T)$ is just the set of fixed points of $S$ as a mapping of $F(T)$. By 7.1.6, this set is compact, convex and non-empty. □

LEMMA 7.1.8   *Any finite intersection of sets $F(T)$ is compact, convex and non-empty.*

*Proof.* Induction, using essentially the argument of 7.1.7.  □

*Proof of Theorem* 7.1.5   Since any finite intersection of sets $F(T)$ is non-empty, and since $\mathcal{M}$ is compact, the intersection of all the sets $F(T)$ is non-empty.  □

*Direct proof of* 7.1.5. We assume, without loss of generality, that $\mathcal{A}$ is a semigroup and contains, with each $U$, the operator $U^{(n)}$ defined by

$$U^{(n)}x = (x + Ux + U^2x + \ldots + U^{n-1}x)/n.$$

(Otherwise, a sequence of extensions of $\mathcal{A}$ gives a larger family of mappings with the required properties.) For $U, V \in \mathcal{A}$, we have

$$(UV)\mathcal{M} = U(V\mathcal{M}) \subset U\mathcal{M}.$$

Hence $V\mathcal{M} \cap U\mathcal{M} \supset (UV)\mathcal{M} \neq \varnothing$. By induction any finite intersection of sets $U\mathcal{M}$ is non-empty. Since $U$ is continuous, each set $U\mathcal{M}$ is compact. Thus by the compactness of $\mathcal{M}$ the intersection

$$\mathcal{L} = \bigcap_{U \in \mathcal{A}} U\mathcal{M}$$

is non-empty. Let $z \in \mathcal{L}$. We will show that $z$ is a fixed point for each operator $U$ in $\mathcal{A}$. In fact, for arbitrarily large $n$, we can find $y$ in $\mathcal{M}$ such that $z = U^{(n)}y$. Thus,

$$Uz - z = n^{-1}(U^n y - y) \in n^{-1}\mathcal{M}_1$$

where $\mathcal{M}_1 = \{x - y : x \in \mathcal{M}, y \in \mathcal{M}\}$. Since $\mathcal{M}$ is compact, $\mathcal{M}_1$ is compact and hence bounded. Thus $\bigcap_n n^{-1}\mathcal{M}_1 = \{0\}$ so that $Uz - z = 0$.  □

THEOREM 7.1.9   *In 7.1.5 we can allow one mapping $T_0$ to be non-affine; there will still be a common fixed point for all the mappings $T$ in $\mathcal{A}$.*

*Proof.* Let $\mathcal{B} = \mathcal{A} - \{T_0\}$. The set $\mathcal{X}$ of common fixed points of the mappings $T$ in $\mathcal{B}$ is compact, convex and non-empty by 7.1.5, and is mapped into itself by $T_0$. Thus $T_0$ has a fixed point in $\mathcal{X}$, and this is a fixed point for all $T$ in $\mathcal{A}$.  □

We now give some theorems concerning families of mappings which need not be affine. Such theorems are known for several classes of non-expansive operators; except for 7.1.11, they depend on special properties of the fixed point sets of these operators.

**THEOREM 7.1.10** *Let $\mathcal{M}$ be a complete metric space and $\{T_i\}$ a commuting family of mappings of $\mathcal{M}$ into $\mathcal{M}$. If one mapping, say $T_0$, is a contraction mapping, then there is a common fixed point for the family.*

*Proof.* By 7.1.3, the fixed point for $T_0$ is a fixed point for each $T_i$. □

**THEOREM 7.1.11** (*DeMarr*)   *Let $\mathcal{M}$ be a compact convex subset of a normed space. Let $\mathcal{F}$ be a family of commuting non-expansive mappings of $\mathcal{M}$ into $\mathcal{M}$. Then there is a common fixed point for the family $\mathcal{F}$.*

*Proof.* See DeMarr (1963 *a*). □

We will now drop the compactness assumption. We need the following interesting geometric property of a non-expansive operator (related to the fact that isometric Banach space operators must be affine: see Banach (1932) for this).

**LEMMA 7.1.12** *Let $\mathcal{M}$ be a convex subset of a strictly convex space $\mathcal{B}$. If $T$ is a non-expansive mapping of $\mathcal{M}$ into $\mathcal{B}$ then the set of fixed points of $T$ is convex.*

*Proof.* Let $a$ and $b$ be fixed points of $T$. Let $c$ lie on the line segment joining $a$ to $b$. Then we have

$$\|Ta-Tc\| \leqslant \|a-c\|, \quad \|Tb-Tc\| \leqslant \|b-c\|. \tag{1}$$

If either inequality (1) were strict we would have

$$\|a-b\| = \|Ta-Tb\| \leqslant \|Ta-Tc\| + \|Tb-Tc\|$$
$$< \|a-c\| + \|b-c\| = \|a-b\|,$$

a contradiction; thus in (1) we have equality. It follows that

$$\|a-Tc\| = \|a-c\|, \quad \|b-Tc\| = \|b-c\|.$$

Since the space is strictly convex, the point $c$ is characterised by its distances from $a$ and $b$; thus $Tc = c$. □

REMARK 7.1.13 Lemma 7.1.12 cannot be extended to arbitrary Banach spaces. The following example is given by DeMarr. Consider the space $R^2$ with the norm

$$\|(x,y)\| = \sup\{|x|, |y|\}$$

and the non-expansive mapping $(x,y) \to (|y|, y)$ of $R^2$ into $R^2$. The fixed point set is the union of the diagonals in the first and fourth quadrants of the plane.

LEMMA 7.1.14 (*Mazur*)   *Any convex strongly closed subset of a Banach space is also weakly closed.*

*Proof.* See for instance Day (1958).   □

THEOREM 7.1.15 (*Browder, 1965b*)   *Let $\mathscr{M}$ be a bounded closed convex subset of a Hilbert space $\mathscr{H}$. Let $\mathscr{F}$ be a commuting family of non-expansive mappings of $\mathscr{M}$ into $\mathscr{M}$. Then the mappings in $\mathscr{F}$ have a common fixed point.*

(The result extends, by the same argument, to uniformly convex spaces; further possible extensions are to the cases listed in 5.1.4.)

*Proof.* For each $T$ in $\mathscr{F}$, $F(T)$ is non-empty (by 5.1.3), convex (by 7.1.12) and strongly closed. By 7.1.14, $F(T)$ is weakly closed. As in the proof of 7.1.8 we see that any finite intersection of sets $F(T)$ is non-empty and weakly compact. Since $\mathscr{M}$ is weakly compact, the intersection of all sets $F(T)$ is non-empty. Thus there exists a common fixed point.   □

## 7.2 'Downward Induction'

By this phrase we mean an argument in which we look for a minimal invariant set, show that it can contain only one point, and conclude that this must be a fixed point. This type of argument has been used to prove various fixed point theorems (see in particular DeMarr (1963a) and §7.3). We give one version of the argument.

THEOREM 7.2.1 (*Downward Induction Theorem*)   *Suppose that*

   (i)   *there is a non-empty compact (convex) set $\mathscr{M}_0$, invariant under a family of operators $\mathscr{F}$;*

(ii) *if $\mathcal{M}_1$ is any compact (convex) set invariant under $\mathcal{F}$ and if $\mathcal{M}_1$ has more than one point then $\mathcal{M}_1$ contains a strictly smaller compact (convex) invariant set.*

*Then there is a common fixed point for $\mathcal{F}$.*

*Proof.* A minimal compact (convex) invariant subset of $\mathcal{M}_0$ exists by Zorn's lemma and must consist of a single point. This point is thus fixed for all operators in $\mathcal{F}$. □

As an illustration we give

**THEOREM 7.2.2** *Let $\mathcal{F}$ be a commuting family of continuous mappings $T_\alpha$ of a compact set $\mathcal{M}$ into $\mathcal{M}$. Assume that for any compact invariant subset $\mathcal{M}_1$ of $\mathcal{M}$ (with more than one point) there is a mapping $T_\alpha$ in $F$ such that*

$$\operatorname{diam}(T_\alpha \mathcal{M}_1) < \operatorname{diam} \mathcal{M}_1. \qquad (*)$$

*Then there is a common fixed point for $\mathcal{F}$.*

*Proof.* If $\mathcal{M}_1$ is any compact invariant subset, choose $T_\alpha$ to give (*); then $T_\alpha \mathcal{M}_1$ is a smaller compact invariant subset. By the downward induction theorem, there must be a common fixed point. □

**REMARK** Theorem 5.2.3 is a special case of 7.2.2.

## 7.3  Groups and semigroups of mappings

**THEOREM 7.3.1** (*Kakutani*, 1938) *Let $\mathscr{V}$ be a locally convex space, $\mathscr{N}$ a subset of $\mathscr{V}$ and $\mathscr{G}$ a **group** of affine mappings of $\mathscr{N}$ into $\mathscr{N}$. Assume that (using the **same topology** throughout)*

 (i)  *$\mathcal{M}$ is a non-empty compact convex subset of $\mathscr{N}$;*
 (ii) *each mapping $T$ in $\mathscr{G}$ is continuous on $\mathcal{M}$ and $T\mathcal{M} \subset \mathcal{M}$;*
 (iii) *the mappings in $\mathscr{G}$ are equicontinuous on $\mathcal{M}$.*

*Then there is a common fixed point for $\mathscr{G}$.*

**REMARK** In applications $\mathscr{V}$ is usually a Banach space with either the norm topology or the weak topology.

**THEOREM 7.3.2** (*Ryll-Nardzewski*, 1966) *Let $\mathscr{V}$ be a normed space, $\mathscr{N}$ a subset of $\mathscr{V}$ and $\mathscr{G}$ a **semigroup** of affine mappings of $\mathscr{N}$ into $\mathscr{N}$. Assume that* (i)′ $\mathcal{M}$ is a non-empty subset of $\mathcal{N}$, compact in the **weak topology**;

(ii)′ each mapping $T$ in $\mathcal{G}$ is **norm-continuous** on $\mathcal{M}$ and $T\mathcal{M} \subset \mathcal{M}$;

(iii)′ $\mathcal{G}$ is distal (that is, for $x \neq y$, 0 is not in the **norm closure** of the set $\{Tx - Ty : T \in \mathcal{G}\}$).

Then there is a common fixed point for $\mathcal{G}$.

R EMARKS   (a) Ryll-Nardzewski (1966), Namioka and Asplund (1967) and Greenleaf (1969) prove a more general form of 7.3.2 in which the norm topology is replaced by any locally convex topology.

(b) For a group $\mathcal{G}$, condition (iii) (in the norm topology) implies (iii)′. Thus the norm topology version of 7.3.1 follows from 7.3.2, and can be generalised to a semigroup if we replace (iii) by (iii)′.

(c) In a reflexive space we can often prove compactness of $\mathcal{M}$ by using the weak topology. In such a case, 7.3.2 is likely to be easier to apply than 7.3.1, since (iii)′ is more easily checked than (iii). Often the operators are isometries and (iii)′ is trivial while (iii) (in the weak topology) is awkward or false.

*Proof of theorem* 7.3.1. For the general case the proof was sketched by Kakutani (1938); see Dunford and Schwartz (1958) for the details. We assume that $\mathcal{V}$ *is a Banach space with the norm topology and the operators* $T$ *are isometries of* $\mathcal{M}$ *onto* $\mathcal{M}$; most applications to Banach spaces can be reduced to this case by renorming the space. The result follows from 7.3.6 below by the downward induction theorem 7.2.1. $\square$

L EMMA 7.3.3   *If $\mathcal{M}$ is a compact convex set containing more than one point then $\mathcal{M}$ contains a non-diametral point $P$ (that is, $\mathcal{M} \subset N(p, d)$ with $d < \operatorname{diam}\mathcal{M}$).*

*Proof.* Let $\operatorname{diam}\mathcal{M} = 2\Delta > 0$. For a suitable integer $n$ and suitable points $k_1, ..., k_n$ in $\mathcal{M}$, $\mathcal{M}$ is covered by the open neighbourhoods $N(k_1, \Delta), ..., N(k_n, \Delta)$. Let $p = n^{-1}(k_1 + ... + k_n)$. For any $m \in \mathcal{M}$ we have $\|k_i - m\| < \Delta$ for some $i$ and $\|k_i - m\| \leqslant 2\Delta$ for all $i$; thus

$$\|p - m\| < n^{-1}(\Delta + 2(n-1)\Delta) = (2 - n^{-1})\Delta \quad (m \in \mathcal{M}).$$

Thus we can take $d = (2 - n^{-1})\Delta$. $\square$

**Lemma** 7.3.4 *If $T$ is an isometry and $m$ is in the domain $\mathscr{D}(T)$ then*
$$T(N(m,\delta) \cap \mathscr{D}(T)) \subset N(Tm,\delta).$$

*Proof.* Obvious. $\square$

**Lemma** 7.3.5 *If a set $\mathscr{M}$ is invariant under a group of mappings $\mathscr{G}$ then each $T$ in $\mathscr{G}$ maps $\mathscr{M}$ onto $\mathscr{M}$.*

*Proof.* $TT^{-1}\mathscr{M} = I\mathscr{M} = \mathscr{M}$ requires that $T$ be 'onto'. $\square$

**Lemma** 7.3.6 *If $\mathscr{M}$ satisfies* (i), (ii) *and* (iii) *and has more than one point then $\mathscr{M}$ contains a strictly smaller set $\mathscr{K}$ satisfying* (i), (ii) *and* (iii).

*Proof.* Choose $p$ and $d$ as in 7.3.3. Then
$$\mathscr{K} = \mathscr{M} \cap \bigcap_{m \in \mathscr{M}} \bar{N}(m,d) \neq \varnothing$$
(since $p \in \mathscr{K}$). Since
$$T\mathscr{K} \subset T\mathscr{M} \cap \bigcap_{m \in \mathscr{M}} \bar{N}(Tm,d) \quad \text{(by 7.3.4)}$$
$$= \mathscr{M} \cap \bigcap_{u \in \mathscr{M}} \bar{N}(u,d) \quad \text{(by 7.3.5)}$$
$$= \mathscr{K} \quad (T \in \mathscr{G}),$$

$\mathscr{K}$ is invariant under $\mathscr{G}$. From its definition, $\mathscr{K}$ is compact convex and non-empty. Clearly, $\operatorname{diam}\mathscr{K} \leqslant d < \operatorname{diam}\mathscr{M}$. $\square$

*Sketch of proof of theorem* 7.3.2. This follows from 7.3.8 below by downward induction. $\square$

**Lemma** 7.3.7 *Given a separable Banach space $\mathscr{V}$, a convex weakly compact subset $\mathscr{M}$ and a number $\delta > 0$, $\mathscr{M}$ has a closed convex proper subset $\mathscr{C}$ with $\operatorname{diam}(\mathscr{M} - \mathscr{C}) < \frac{1}{2}\delta$.*

*Proof.* This can be derived from the Krein–Milman theorem; see Greenleaf (1969). $\square$

**Lemma** 7.3.8 *If $\mathscr{M}$ satisfies* (i)′, (ii)′ *and* (iii)′ *and has more than one point, $\mathscr{M}$ contains a strictly smaller set $\mathscr{K}$ satisfying* (i)′, (ii)′ *and* (iii)′.

*Proof.* We can assume (see Greenleaf (1969)) that $\mathscr{V}$ is separable. Choose $x$ and $y$ in $\mathscr{M}$ with $x \neq y$, and put
$$\inf\{\|Ty - Tx\| : T \in \mathscr{G}\} = \delta > 0.$$

Let $\mathscr{C}$ be the set given by 7.3.7. If $T \in \mathscr{G}$, we cannot have both $Tx$ and $T(\frac{1}{2}x + \frac{1}{2}y)$ in $\mathscr{M} - \mathscr{C}$; for the distance between these points is too great. Thus $T(\frac{1}{2}x + \frac{1}{2}y)$ must be in $\mathscr{C}$ (for otherwise $Tx$ and similarly $Ty$ would be in $\mathscr{C}$; contradiction). Thus if we set

$$\mathscr{K} = \overline{\text{co}}\,\{T(\tfrac{1}{2}x + \tfrac{1}{2}y) : T \in \mathscr{G}\}$$

we have $\mathscr{K} \subset \mathscr{C}$, so that $\mathscr{K}$ satisfies the required conditions. (We use the norm-closed convex cover; by 7.1.14 this is weakly closed, so weakly compact.) $\square$

REMARK   If $\mathscr{V}$ is a Hilbert space, 7.3.7 can be proved by a simple argument, without assuming separability. The proof of 7.3.8 sketched above is thus a complete proof in this case.

## Exercises

1. If $\mathscr{A}$ is a commuting family of mappings of a set $\mathscr{M}$ into $\mathscr{M}$, such that $\bigcap_{T \in \mathscr{A}} T\mathscr{M}$ is a one-point set, then there is a common fixed point for $\mathscr{A}$.

2. (*Folkman*) (*a*) If $S$ and $T$ are commuting continuous mappings of a metric space $\mathscr{M}$ into $\mathscr{M}$, if $x$ is a fixed point for $S$ and if $y = \lim T^n x$ exist, then $y$ is a common fixed point for $S$ and $T$.

(*b*) If $S$ and $T$ are commuting continuous mappings of $[0, 1]$ into $[0, 1]$, and $T$ is monotone, there is a common fixed point for $S$ and $T$.

3. Show that an operator is affine if and only if its graph is convex.

4. Show that an affine mapping is continuous on each line segment in its domain.

5. Extend 7.1.10 to the case where $T_0^K$ is a contraction mapping for some integer $K > 1$.

6. Give an example of a group of continuous mappings of $[-1, 1]$ into itself, with no common fixed point.

7. Show that a semigroup containing a contraction mapping is not distal.

8. Let $\mathscr{M}$ be a closed convex subset of a Banach space. Let $\{T_\alpha\}$ be a commuting family of continuous affine mappings of $\mathscr{M}$ into $\mathscr{M}$. Suppose that $T_\alpha\mathscr{M}$ is precompact for one value of $\alpha$. Show that there is a common fixed point for the family.

# 8. *Existence of invariant means*

We will use fixed point theorems (mainly those of Chapter 7) to establish a number of related results. The method is best illustrated by § 8.1. In the rest of the chapter we have to use the weak * topology on the dual of a Banach space; the arguments are thus relatively complicated. We consider only real functions and sequences.

A converse method exists; from amenability (the existence of a left-invariant mean for a semigroup) follow fixed point properties of realisations of the semigroup (as a semigroup of mappings). See § 8.4.

## 8.1 Almost periodic functions

One of the basic facts about an almost periodic function on $(-\infty, \infty)$ is the existence of a mean value (which can be used in defining various norms). This fact can be obtained fairly easily from the following theorem (the value of the constant function gives the mean value for $f$).

DEFINITION 8.1.1 A bounded function $f$ on a group $\mathscr{G}$ is called (*left-*) (*uniformly*) *almost periodic* if the set

$$\mathscr{M}_f = \{f_\alpha : \alpha \in \mathscr{G}\}$$

is precompact in the uniform norm. (Here $f_\alpha$ denotes the function $f_\alpha(x) = f(\alpha x)$.)

THEOREM 8.1.2 *If a bounded function $f$ is almost periodic on a group $\mathscr{G}$ then $\mathscr{M} = \overline{\mathrm{co}}\,(\mathscr{M}_f)$ contains a constant function.*

*Proof.* The operators $T_\alpha$ defined by $T_\alpha g = g_\alpha$ form a group of isometries of the compact set $\mathscr{M}$ into $\mathscr{M}$. The $T_\alpha$ are equicontinuous and therefore, by Kakutani's theorem 7.3.1 have a common fixed point in $\mathscr{M}$. Clearly this fixed point is a constant function. $\square$

We can obtain the same result if the function $f$ is 'weakly almost periodic', i.e. if $\mathcal{M}_f$ is precompact in the weak topology. In this case the Ryll-Nardzewski theorem 7.3.2 must be used in place of 7.3.1.

## 8.2   Banach limits

We first show that a translation-invariant 'limit' function can be defined on the space $(m)$ of bounded sequences.

THEOREM 8.2.1 (*Banach*)   *We can assign to each bounded sequence* $a = (a_1, a_2, \ldots)$ *of a scalars a 'limit'* $L(a)$ *such that*
   (i)   $L$ *is a linear functional on* $(m)$,
   (ii)   $L(1, 1, 1, \ldots) = 1$,
   (iii)   $L(a) \geqslant 0$ *if* $a \geqslant 0$,
   (iv)   $L(a_1, a_2, \ldots) = L(a_2, a_3, \ldots)$.

REMARK 8.2.2   Given (i) we can see that (ii) and (iii) together are equivalent to
   (v)   $\inf a_n \leqslant L(a) \leqslant \sup a_n$,
from which it follows that $\|L\| \leqslant 1$. Also from (v) and (iv) we can deduce
   (vi)   $\liminf a_n \leqslant L(a) \leqslant \limsup a_n$,
so that the 'limit' functional has reasonable properties.

We require the following facts about the weak* topology.

REMINDER 8.2.3   If $\mathcal{B}$ is a Banach space and $\mathcal{B}^*$ its dual,
   (*a*) we can define a topology, the *weak\* topology*, on $\mathcal{B}^*$ as follows. A base of neighbourhoods of a point $U$ in $\mathcal{B}^*$ consists of finite intersections of neighbourhoods of the form
$$N(U, a) = \{L \in \mathcal{B}^* : |L(a) - U(a)| < 1\}$$
(for $a \in \mathcal{B}$);
   (*b*) for each $x \in \mathcal{B}$ the function $\hat{x}$ on $\mathcal{B}^*$, defined by
$$\hat{x}(f) = f(x) \, (f \in \mathcal{B}^*),$$
is continuous in the weak* topology;
   (*c*) the unit ball of $\mathcal{B}^*$ is compact in the weak* topology.
*Proof of* 8.2.1. We consider the set
$$\mathcal{M} = \{L : L \text{ satisfies (i), (ii), (iii)}\}.$$

$\mathcal{M}$ is non-empty: for example take $L(a) = a_1$. Also $\mathcal{M}$ is a subset of the unit ball in $(m)^*$, by 8.2.2, and $\mathcal{M}$ is clearly convex. By (v),

$$\mathcal{M} = \bigcap_a \{L : L(a) \leqslant \sup a_n\} \cap \bigcap_a \{L : L(a) \geqslant \inf a_n\},$$

an intersection of half spaces in $(m)^*$ which are weak*-closed since $a$ defines a weak*-continuous functional on $(m)^*$. Thus $\mathcal{M}$ is a weak*-closed subset of a weak*-compact set, so that: *$\mathcal{M}$ is a weak\*-compact, convex subset of $(m)^*$.*

We now consider a mapping $T$ of $\mathcal{M}$ into $\mathcal{M}$ defined by:

$$(TL)(a_1, a_2, \ldots) = L(a_2, a_3, \ldots).$$

$T$ is weak*-continuous since $T$ is linear and for the neighbourhood $N(U; a)$ we have

$$
\begin{aligned}
T^{-1}(N(U; a)) &= \{L : TL \in N(U; a)\} \\
&= \{L : |TL(a) - U(a)| < 1\} \\
&= \{L : |L(a_2, a_3, \ldots) - U(a_1, a_2, \ldots)| < 1\} \\
&= N(U_1; (a_2, a_3, \ldots)),
\end{aligned}
$$

where $U_1$ is any point of $(m)^*$ satisfying

$$U_1(a_2, a_3, \ldots) = U(a_1, a_2, \ldots).$$

The conditions of Tychonoff's theorem 2.3.11 are thus satisfied and there must exist a fixed point for the mapping $T$. However, by the definition of $T$, this fixed point is a functional $L$ satisfying (iv), as well as (i), (ii) and (iii). $\square$

If we wish to consider functions rather than sequences we must consider a family of translation operators in place of one translation operator.

**THEOREM 8.2.4**  *We can assign a limit $L(f)$ to each bounded function $f$ on $[0, \infty]$ such that*
   (i)  *$L$ is linear,*
   (ii)  *$L(1) = 1$,*
   (iii)  *$L(f) \geqslant 0 \quad (f \geqslant 0)$,*
   (iv)  *$L(f_\alpha) = L(f) \quad (\alpha > 0)$*
*(where $\mathbf{1}(x) \equiv 1$ and $f_\alpha(x) \equiv f(x + \alpha)$).*

*Proof.* Similar to that of 8.2.1. We consider the set

$$\mathcal{M} = \{L : L \text{ satisfies (i), (ii), (iii)}\},$$

and the operators $T_\alpha$ such that $(T_\alpha L)(f) = L(f_\alpha)$. We find that $\mathcal{M}$ is weak*-compact and that the $T_\alpha$ map $\mathcal{M}$ into $\mathcal{M}$. Since the $T_\alpha$ commute we can use the Markov–Kakutani theorem, which yields a common fixed point for the $T_\alpha$ in $\mathcal{M}$; in other words, a functional $L$ satisfying (iv) as well as (i), (ii) and (iii).  □

In precisely the same way we prove that an invariant mean exists on the space of bounded functions on any Abelian semigroup.

THEOREM 8.2.5   *If $\mathcal{G}$ is an Abelian semigroup there exists a functional $L$ on $m(\mathcal{G})$ such that*

(i)   *$L$ is linear,*          (ii)   $L(\mathbf{1}) = 1,$

(iii)   $L(f) \geqslant 0$   $(f \geqslant 0)$,     (iv)   $L(f_\alpha) = L(f)$   $(\alpha \in \mathcal{G})$,

*where $\mathbf{1}(x) \equiv 1$ and $f_\alpha(x) \equiv f(\alpha x)$.*

If $\mathcal{G}$ is not Abelian, a functional $L$ with the properties (i) to (iv) need not exist on $m(\mathcal{G})$, even if $\mathcal{G}$ is a group. If $\mathcal{G} = S^2$ (the group of rotations of a two-sphere), an example due to Hausdorff (1914, p. 469) concerning subsets of the sphere with pathological properties under rotation, shows that no functional $L$ can satisfy (i) to (iv). It was shown by v. Neumann (1929) that: for a functional $L$ satisfying (i) to (iv) to exist, $\mathcal{G}$ must not contain a free group on two generators. See Greenleaf (1969).

We remark that a functional $L$ with properties (i) to (iv) exists on $C(S^2)$; in fact $L(f) = \iint_{S^2} f$. (This is a special case of 8.3.1 below.)

## 8.3   Haar measure

The Haar integral (where it exists) on a topological group $\mathcal{G}$ is defined in the first instance as a function $L$ on $C(\mathcal{G})$ satisfying conditions (i), (iii) and (iv) of 8.2.5, and a normalisation condition. If $\mathcal{G}$ is compact the normalisation condition is condition (ii) of

8.2.5 and we can show that $L$ exists by a method similar to that used in §8.2. Since we are now dealing with a group of translation operators which is not necessarily Abelian, we use Kakutani's theorem 7.3.1 instead of the Markov–Kakutani theorem 7.1.5. The most delicate step is the verification of the equicontinuity condition. We will not discuss the existence of Haar measure on non-compact groups.

(We write $_\beta f_\alpha$ for the function defined by $_\beta f_\alpha(x) = f(\beta x \alpha)$.)

**THEOREM 8.3.1** *If $\mathscr{G}$ is a compact group there exists a functional $L$ on $C(\mathscr{G})$ such that*

   (i)  *$L$ is linear,*
   (ii) *$L(\mathbf{1}) = 1$,*
   (iii) *$L(f) \geqslant 0$ if $f \geqslant 0$,*
   (iv) *$L(_\beta f_\alpha) = L(f)$   $(\alpha, \beta \in G, f \in C(\mathscr{G}))$.*

*Sketch of proof.* Let

$$\mathscr{M} = \{L : L \text{ satisfies (i), (ii), (iii)}\}.$$

By arguments similar to those used for 8.2.1, $\mathscr{M}$ is easily shown to be a weak* closed subset of the unit ball in $C(\mathscr{G})^*$; hence $\mathscr{M}$ is a convex weak*-compact set. The mappings $_\beta T_\alpha$ defined by $(_\beta T_\alpha L)(f) = L(_\beta f_\alpha)$ are linear mappings of $\mathscr{M}$ into $\mathscr{M}$. These mappings form a group. Thus Kakutani's theorem 7.2.1. gives a common fixed point for the $_\beta T_\alpha$; this fixed point is the required functional $L$. The final step requires the following lemma. □

**LEMMA 8.3.2** *The mappings $_\beta T_\alpha$ considered in the above proof are equicontinuous, in the weak* topology on $C(\mathscr{G})$.*

*Sketch of proof.* For details see Dunford and Schwartz (1963, XI.1.1). We must show that for each weak* neighbourhood $N$ of 0 we can find a weak* neighbourhood $M$ of 0 such that

$$L \in M \Rightarrow {}_\beta T_\alpha L \in N \quad (\alpha, \beta \in \mathscr{G}).$$

If $N$ is defined by points $f_1, \dots, f_r$ in $C(\mathscr{G})$, we choose a finite number of functions $g_1, \dots, g_s$ in $C(\mathscr{G})$ such that all translates $_\beta(f_i)_\alpha$ can be approximated within $\epsilon$ (uniformly) by functions $g_j$. The $g_j$ are then used to define the neighbourhood $M$. The possibility of finding suitable functions $g_j$ depends on the lemma below. □

LEMMA 8.3.3   *If f is a continuous function on a compact group $\mathcal{G}$ then the set of translates*

$$\{_\beta f_\alpha : \alpha, \beta \in \mathcal{G}\}$$

*is precompact in the uniform norm.*

*Proof.* Left to the reader. □

## 8.4   Day's fixed point theorem

In the proof of 8.2.4 and 8.2.5 we were concerned with the following situation. We considered an Abelian semigroup $\mathcal{G}$ and a set $\mathcal{M}$ (which can be described as the positive face of the unit sphere in $m(\mathcal{G})^*$). For $\alpha \in \mathcal{G}$ we considered the mapping $T_\alpha$ of $\mathcal{M}$ into $\mathcal{M}$ such that $(T_\alpha L)(f) \equiv L(f_\alpha)$, where $f_\alpha(x) \equiv f(x+\alpha)$. These mappings $T_\alpha$ constitute the regular representation of $\mathcal{G}$. Theorem 8.2.5 thus asserts that the regular representation has a fixed point in $\mathcal{M}$. If $\mathcal{G}$ is not Abelian we define $f_\alpha$ as $f(\alpha x)$ and call the correspondence $\alpha \to T_\alpha$ the left-regular representation of $\mathcal{G}$.

THEOREM 8.4.1 (*Day*)   *Suppose that the left-regular representation of the semigroup $\mathcal{G}$ has a fixed point in $\mathcal{M}$. Then each representation of $\mathcal{G}$ (as a semigroup of affine continuous mappings $S_\alpha$ of a compact convex non-empty set $\mathcal{N}$ into $\mathcal{N}$) has a fixed point.*

*Sketch of proof.* Set up a mapping $\tau$ of $\mathcal{M}$ into $\mathcal{N}$ such that $S_\alpha(\tau x) = \tau(T_\alpha x)$ for all $x$ in $\mathcal{M}$ and all $\alpha$ in $\mathcal{G}$. Then if $x$ is a common fixed point for the $T_\alpha$, $\tau x$ is a common fixed point for the $S_\alpha$. See Day (1961) or Greenleaf (1969) for the details. □

COROLLARY 8.4.2   *A semigroup $\mathcal{G}$ is amenable (that is, there exists a left-invariant mean on $m(\mathcal{G})$) if and only if each representation of $\mathcal{G}$ (by affine continuous mappings on a compact convex non-empty set) has a fixed point.*

*Proof.* As in the discussion of 8.2.4, we see that a left-invariant mean on $m(\mathcal{G})$ is precisely a fixed point (in the set $\mathcal{M}$) for the left-regular representation of $\mathcal{G}$. Thus the result follows from 8.4.1. □

# 9. Fixed point theorems for many-valued mappings

If each point $x$ of a set $\mathscr{M}$ is mapped onto a *set* $U(x)$, we call $U$ a many-valued mapping. This chapter is concerned with the case where each $U(x)$ is compact and convex. Various other cases have been considered. Eilenberg and Montgomery (1946) allow $U(x)$ to be acyclic (homologically trivial); so do Begle (1950) and Górniewicz and Granas (1970). Ky Fan (1961) merely requires $U(x)$ to be compact; but here $U(x)$ must depend continuously on $x$. Smithson (1965) considers cases where $U(x)$ is finite-valued. For various recent contributions see Fleischman (1970) and Smithson (1972).

## 9.1 Kakutani's theorem

DEFINITION 9.1.1   Let $\mathscr{S}$ and $\mathscr{T}$ be subsets of a normed space $\mathscr{B}$. We will say that $U$ *is a K-mapping* of $\mathscr{S}$ into $\mathscr{T}$ if

(i) for each $x$ in $\mathscr{S}$ a compact convex non-empty subset $U(x)$ of $\mathscr{T}$ is defined;

(ii) the graph of $U$,

$$\mathscr{G}(U) = \{\langle x, y \rangle : y \in U(x)\}$$

is closed in $\mathscr{S} \times \mathscr{T}$.

Any continuous mapping of $\mathscr{S}$ into $\mathscr{T}$ can be regarded as a $K$-mapping. The results in this chapter thus generalise earlier theorems about continuous mappings.

Condition (ii) is equivalent to the following condition, described as 'upper semi-continuity' by Kakutani (1941):

(ii)' if $x_n \to x$ in $\mathscr{S}$, $y_n \in U(x_n)$ and $y_n \to y$ then $y \in U(x)$.

DEFINITION 9.1.2   A *fixed point* for a $K$-mapping $U$ is a point $x$ such that $x \in U(x)$.

DEFINITION 9.1.3 We say that a subset $\mathscr{S}$ of a normed space has *the Kakutani property* if each $K$-mapping of $\mathscr{S}$ into $\mathscr{S}$ has a fixed point.

THEOREM 9.1.4 *If $\mathscr{S}$ has the Kakutani property then any retract of $\mathscr{S}$ has the Kakutani property.*

*Proof.* Let $r$ be a retraction mapping of $\mathscr{S}$ onto $\mathscr{R}$. Let $U$ be a $K$-mapping of $\mathscr{R}$ into $\mathscr{R}$. Then we define a $K$-mapping $V$ of $\mathscr{S}$ into $\mathscr{S}$ (see exercise 1) by

$$V(x) = U(rx).$$

$V$ has a fixed point $y$ so that $y \in U(ry) \subset \mathscr{R}$. Thus $y = ry$ so that $y \in U(y)$. $\square$

THEOREM 9.1.5 *If $\mathscr{S}$ has the Kakutani property and $\mathscr{S}$ is homeomorphic to $\mathscr{R}$ under an affine mapping then $\mathscr{R}$ has the Kakutani property.*

*Proof.* Left to the reader. $\square$

THEOREM 9.1.6 (*Kakutani*, 1941) *Any compact convex nonempty subset of $R^m$ has the Kakutani property.*

*Proof.* If the result is proved for simplexes, the general result will follow by 9.1.4. Assume then that $\mathscr{S}$ is a closed $r$-simplex. If $\epsilon_n > 0$ we can construct an $\epsilon_n$-fine simplicial subdivision of $\mathscr{S}$. We can then define a mapping $T_n$ of $\mathscr{S}$ into $\mathscr{S}$ such that

(1) $T_n(x_i^n) \in U(x_i^n)$ for each vertex $x_i^n$ of the subdivision;

(2) $T_n$ is affine on each simplex of the subdivision.

$T_n$ is then a continuous mapping of $\mathscr{S}$ into $\mathscr{S}$ and so has a fixed point $z_n$ by 2.1.11: that is, $T_n z_n = z_n$. We choose our numbering of the vertices so that

$$z_n \in \mathrm{co}\,(x_0^n, x_1^n, ..., x_r^n).$$

Thus $\qquad z_n = \Sigma c_i^n x_i^n$ where $c_i^n \geqslant 0$ and $\Sigma c_i^n = 1.$

(We write $\Sigma$ for a sum from $i = 0$ to $i = r$, throughout the argument.) We will take a sequence $\epsilon_n \to 0$. We can assume without loss of generality that as $n \to \infty$ we have $z_n \to z \in \mathscr{S}$ so that also $x_i^n \to z$ for $0 \leqslant i \leqslant r$. We can also assume that for each $i$ ($0 \leqslant i \leqslant r$), $T_n x_i^n$ converges, say to $y_i$; and that $c_i^n$ converges, say to $c_i$. Clearly $c_i \geqslant 0$ and $\Sigma c_i = 1$. Since $\langle x_i^n, T_n x_i^n \rangle$ is in $\mathcal{G}(U)$ and converges to $\langle z, y_i \rangle$, this point is in $\mathcal{G}(U)$, i.e. $y_i \in U(z)$.

Thus $z = \lim z_n = \lim T_n z_n = \lim \Sigma c_i^n T_n x_i^n = \Sigma c_i y_i \in U(z)$, since $U(z)$ is convex. $\square$

## 9.2 Generalisations

Kakutani's theorem was extended to Banach spaces by Bohnenblust and Karlin(1950) (by a method similar to Schauder's proof of 2.3.7) and to locally convex spaces by Ky Fan (1952) and Glicksberg (1952). We give a discussion of the Banach space case which generalises our proof of Schauder's theorem. This is the case needed for an application (Lasota and Opial, 1965) to the differential problem $dx/dt - A(t)\, x \in F(t, x)$, where $F$ is set-valued.

**THEOREM 9.2.1** *The Hilbert cube $\mathcal{H}_0$ has the Kakutani property.*

*Proof.* We consider the projections $P_n$ in $l^2$:

$$P_n(x_1, x_2, \ldots) = (x_1, x_2, \ldots, x_n, 0, 0, \ldots).$$

The theorem follows from the following lemma and the properties of $P_n$ discussed in 2.3.5. $\square$

**LEMMA** *Let $\mathcal{X}$ be a compact convex subset of a normed space. Suppose that for $n = 1, 2, \ldots$ there is a continuous linear mapping $P_n$ of $\mathcal{X}$ into $\mathcal{X}$ such that*

(i) $\|P_n x - x\| < n^{-1}$ $(x \in \mathcal{X})$;

(ii) $P_n \mathcal{X}$ *has the Kakutani property.*

*Then $\mathcal{X}$ has the Kakutani property.*

*Proof.* Let $U$ be a $K$-mapping of $\mathcal{X}$ into $\mathcal{X}$. It is easily seen (exercise 2) that $P_n U$ is a $K$-mapping of $P_n \mathcal{X}$ into $P_n \mathcal{X}$. Thus $P_n U$ has a fixed point $y_n$ in $P_n \mathcal{X}$,

$$y_n \in P_n U(y_n).$$

Thus $y_n = P_n z_n$ with $z_n \in U(y_n)$. Assume (without loss of generality) that $y_n \to y \in \mathcal{X}$. Since $\|y_n - z_n\| = \|P_n z_n - z_n\| \to 0$, we see that $z_n \to y$. Thus the points $\langle y_n, z_n \rangle$ in $\mathcal{G}(U)$ converge to $\langle y, y \rangle$. Since $\mathcal{G}(U)$ is closed, $\langle y, y \rangle$ is in $\mathcal{G}(U)$; that is, $y$ is a fixed point for $U$. $\square$

THEOREM 9.2.2 *Any compact convex non-empty subset $\mathscr{X}$ of a normed space has the Kakutani property.*

*Proof.* By 2.3.3, $\mathscr{X}$ is homeomorphic, under a linear mapping, to a compact convex subset $\mathscr{Y}$ of the Hilbert cube $\mathscr{H}_0$. By 2.1.4, $\mathscr{Y}$ is a retract of $\mathscr{H}_0$. Since $\mathscr{H}_0$ has the Kakutani property, 9.1.4 and 9.1.5 show that $\mathscr{Y}$ and $\mathscr{X}$ have this property. $\square$

THEOREM 9.2.3 *Let $\mathscr{M}$ be a closed convex subset of a Banach space and let $T$ be a $K$-mapping of $\mathscr{M}$ into a compact subset of $\mathscr{M}$. Then $T$ has a fixed point.*

*Proof.* $T$ gives a $K$-mapping of the compact convex set $\overline{\mathrm{co}}(T\mathscr{M})$ into itself; a fixed point exists by 9.2.2. $\square$

THEOREM 9.2.4 *Let $\mathscr{M}$ be a closed convex subset of a Banach space $\mathscr{B}$, with boundary $\partial\mathscr{M}$. Let $U$ be a $K$-mapping of $\mathscr{M}$ into a compact subset of $\mathscr{B}$ such that $U(x) \subset \mathscr{M}$ for $x \in \partial\mathscr{M}$. Then $U$ has a fixed point.*

*Proof.* If the interior $\mathscr{M}^0$ is empty, $\mathscr{M} = \partial\mathscr{M}$ and theorem 9.2.3 gives the result. If $\mathscr{M}^0 \neq \varnothing$, assume without loss of generality that $0 \in \mathscr{M}^0$. By the compactness assumption, the range of $U$, $\bigcup_{x \in \mathscr{M}} U(x)$, is contained in $c\mathscr{M}$ for some $c \geqslant 1$. Fix this $c$, and define a $K$-mapping $V$ of $c\mathscr{M}$ into $c\mathscr{M}$ by

$$V(x) = U(rx),$$

where $r$ is the radial retraction onto $\mathscr{M}$ defined by

$$rx = x/\max(g(x), 1)$$

(see the proof of 4.2.4; $g$ is the Minkowski functional of $\mathscr{M}$). Now $V$ has a fixed point $z$ by 9.2.3. If we had $z \notin \mathscr{M}$ this would give $rz \in \partial\mathscr{M}$ and $V(z) = U(rz) \subset \mathscr{M}$; thus $z$ could not be a fixed point. Thus the fixed point $z$ is in $\mathscr{M}$ and hence

$$z \in V(z) = U(z). \quad \square$$

We can also give a continuation theorem for $K$-mappings.

THEOREM 9.2.5 *Theorem 4.3.3 remains true if $U$ is a $K$-mapping.*

*Proof.* Modify the proof of 4.3.3 (see Smart, (1973$b$)). $\square$

Granas (1959) contains some other theorems on $K$-mappings.

### 9.3   Theory of games

Kakutani gave 9.1.6 as a means of proving the fundamental theorem of the theory of games. We will sketch this application. For applications of 9.1.6 to problems in economics, see Karlin (1959, §8.7).

We consider games which can be reduced to the following form. $\mathscr{M}$ is a compact convex subset of $R^m$; $\mathscr{N}$ is a compact convex subset of $R^n$. A real function $f(x, y)$ is defined for $x \in \mathscr{M}$, $y \in \mathscr{N}$ and is linear in $x$ and in $y$. Player $P$, who wishes to maximise $f(x, y)$, chooses $x$ from $\mathscr{M}$; we call $x$ his 'strategy'. Player $Q$, who wishes to minimise $f(x, y)$, chooses $y$ from $\mathscr{N}$; $y$ is his 'strategy'.

In this setting we can enquire: does there exist a pair of strategies $\langle X, Y \rangle$ such that

(1) $f(x, Y) \leqslant f(X, Y) \leqslant f(X, y)$    $(\forall x \in \mathscr{M}, \forall y \in \mathscr{N})$.

If such a pair of strategies exists, there is no incentive for either player to change from these strategies.

The relations (1) can be written as

(2) $f(X, Y) = \inf_y f(X, y) = \sup_x f(x, Y)$;

throughout this section we assume that $x$ varies over $\mathscr{M}$ and $y$ over $\mathscr{N}$. Thus $\langle X, Y \rangle$ would be given by a fixed point for a mapping defined as follows: for each $y$, player $P$ chooses $x(y)$ such that

(3) $f(x(y), y) = \sup_x f(x, y)$

and for each $x$, player $Q$ chooses $y(x)$ such that

(4) $f(x, y(x)) = \inf_y f(x, y)$.

(Thus, given strategies $x$ and $y$, each player chooses a good answer to the other's strategy.) The mapping

(5) $T : \langle x, y \rangle \rightarrow \langle x(y), y(x) \rangle$

of $\mathscr{M} \times \mathscr{N}$ into itself is the one required. This mapping, because of the choices involved, is likely to be discontinuous so that we cannot appeal to Brouwer's theorem for the existence of a fixed point $\langle X, Y \rangle$. Von Neumann (1935) approximated $T$ (not uniformly, of course!) by continuous mappings $T_n$ obtained by a local averaging process. Fixed points for the $T_n$ exist by Brouwer and yield, by compactness, a solution of (2). This procedure involves some technical complications which are avoided by Kakutani's procedure. In this we consider the point-to-set mapping

(6)  $U: \langle x, y \rangle \to \{\langle x(y), y(x) \rangle : (3) \text{ and } (4) \text{ are true}\};$

it is easily seen that this is a $K$-mapping. A fixed point for $U$ exists by 9.1.6. This is a point $\langle X, Y \rangle$ satisfying (2), i.e. (1).

This proves the following result, known as the 'saddle-point' theorem or the 'fundamental theorem of the theory of games'.

THEOREM 9.3.1   *If $\mathcal{M}$ and $\mathcal{N}$ are compact convex subsets of Euclidean spaces and $f$ a bilinear real function on $\mathcal{M} \times \mathcal{N}$, there exists a point $\langle X, Y \rangle$ in $\mathcal{M} \times \mathcal{N}$ satisfying (2).*

COROLLARY 9.3.2 (*Minimax theorem*)   *Under the conditions of the theorem, the point $\langle X, Y \rangle$ satisfies*

$$f(X, Y) = \inf_y \sup_x f(x, y)$$
$$= \sup_x \inf_y f(x, y).$$

*Proof.* Write

$$g(x) = \inf_y f(x, y) \quad \text{and} \quad h(y) = \sup_x f(x, y).$$

Clearly, for all $x, y$ we have

$$g(x) \leqslant f(x, y) \leqslant h(y)$$

so that                              $$\sup_x g(x) \leqslant \inf_y h(y).$$

However, for the point $\langle X, Y \rangle$ we have

$$g(X) = f(X, Y) = h(Y)$$

so that

$$\sup_x g(x) \geqslant g(X) = f(X, Y) = h(Y) \geqslant \inf_y h(y)$$
$$\geqslant \sup_x g(x),$$

so that all these inequalities are equalities.  □

## Exercises

We write $\mathcal{L}, \mathcal{M}, \mathcal{N}$ for subsets of a normed space.

1. If $U$ is a $K$-mapping of $\mathcal{M}$ into $\mathcal{N}$ and $T$ a continuous mapping of $\mathcal{L}$ into $\mathcal{M}$, show that $UT$ is a $K$-mapping of $\mathcal{L}$ into $\mathcal{N}$.

2. Let $\mathcal{M}$ be compact. If $U$ is a $K$-mapping of $\mathcal{L}$ into $\mathcal{M}$ and $T$ is a continuous affine mapping of $\mathcal{M}$ into $\mathcal{N}$, show that $TU$ is a $K$-mapping of $\mathcal{L}$ into $\mathcal{N}$.

3. Use the theorems of §9.2 to prove generalisations of 9.3.1. (See Bohnenblust and Karlin (1950).)

4. Define the product of two $K$-mappings by

$$UV(x) = \bigcup_{v \in V(x)} U(v).$$

Show that the product of two affine $K$-mappings is an affine $K$-mapping.

5. Let $\mathcal{V}$ be the open unit ball in $R^n$ and $U$ a $K$-mapping of $\mathcal{V}$ into $\mathcal{V}$. Show that for each $\epsilon > 0$ there is a point $x(\epsilon)$ such that

$$\rho(x(\epsilon),\ U(x(\epsilon))) < \epsilon \quad \text{(consider } (1-\epsilon)\,U).$$

## Problems

1. If a subset $\mathcal{X}$ of a normed space has the fixed point property, must $\mathcal{X}$ have the Kakutani property? (The converse is obvious.)

2. Extend 9.2.3 and (hence) 9.2.4 to normed spaces.

3. Discuss the existence of common fixed points for families of affine $K$-mappings (see exercise 4).

# 10. *Some numerical invariants*

A variety of different (but related) numerical invariants has been used in fixed point theory. We have developed fixed point theory without using these invariants; however, it seems important to give an introduction to some of these invariants and their uses.

## 10.1   The rotation of a vector field

Let $v$ be a continuous field of plane vectors defined on a subset $\mathscr{D}$ of the plane. If $C$ is a simple closed curve in $\mathscr{D}$ such that $v(x) \neq 0$ for $x$ on $C$, the rotation of $v$ on $C$ can be intuitively defined: $\operatorname{rot}(v, C)$ is the number of anticlockwise revolutions made by $v(x)$ when $x$ passes anticlockwise around $C$. It is clear that $\operatorname{rot}(v, C)$ is an integer which is unaltered when $C$ is continuously deformed (provided that $v$ remains non-zero). Thus, if $C$ can be deformed into $C_1$ in $\mathscr{D}$ and if $\operatorname{rot}(v, C) \neq \operatorname{rot}(v, C_1)$, there must be a zero of $v$ between $C$ and $C_1$. In particular, if $C$ can be contracted to a point in $\mathscr{D}$ and if $\operatorname{rot}(v, C) \neq 0$, then there must be a zero of $v$ inside $C$.

We mention three applications. For details of the first two, see Courant and Robbins (1961); for the third, see Brouwer (1952).

THEOREM 10.1.1   *Fundamental theorem of algebra. If $p$ is a complex polynomial of degree $n \geqslant 1$ then $p$ has a complex root.*

*Sketch of proof.* Consider the vector field $v = p$ on the complex plane $\mathscr{D}$. If $C$ is a large circle about 0, we have $\operatorname{rot}(v, C) = \pm n$. Since $C$ can be deformed to a point in $\mathscr{D}$, $v$ has a zero inside $C$. □

THEOREM 10.1.2   *Brouwer's theorem for $B^2$. Any continuous mapping $T$ of the closed unit disc into itself has a fixed point.*

*Sketch of proof.* Consider the vector field $v(x) = Tx - x$ on the disc $\mathscr{D}$ with boundary $C$. If $T$ has no fixed point on $C$ then $\operatorname{rot}(v, C) = 1$. Since $C$ can be deformed to a point in $\mathscr{D}$, $v$ has a zero; that is, $T$ has a fixed point. □

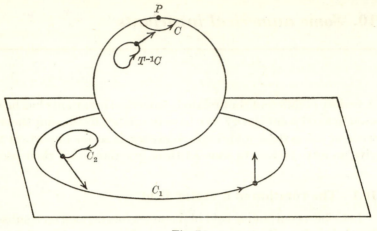

Fig. 7

**COROLLARY**  *Any continuous mapping of $S^2$ into a proper subset of $S^2$ has a fixed point.*

**THEOREM 10.1.3**  *Brouwer's theorem for $S^2$. Any continuous one–one orientation-preserving mapping $T$ of $S^2$ into $S^2$ has a fixed point.*

*Sketch of proof.* See figure 7. We can assume that $T$ is 'onto', and that the north pole $P$ of $S^2$ is not a fixed point (otherwise, the theorem is trivial). Let $C$ be a small circle about $P$. Map $S^2 - \{P\}$ onto a plane. Let $T'$, $C_1$ and $C_2$ be the images in the plane of $T$, $C$, and $T^{-1}C$. Since

$$\mathrm{rot}\,(T' - I, C_1) = 1 \quad \text{and} \quad \mathrm{rot}\,(T' - I, C_2) = -1,$$

there is a zero of $T' - I$ between $C_1$ and $C_2$; this gives a fixed point of $T'$ and hence of $T$. $\square$

To generalise these ideas to higher dimensions we must define the rotation, which is no longer an intuitive concept. For one definition see Alexandroff and Hopf (1935). Here the rotation of $f$ about $S^{n-1}$ is obtained (in simple cases) by considering any ray from 0. The algebraic number of intersections of this ray with $f(S^{n-1})$ is defined once we have oriented $f(S^{n-1})$ in accordance with the orientation of $S^{n-1}$. This algebraic number of intersections is the rotation.

The rotation about a sphere in $R^n$ can also be defined in terms of the degree (§ 10.2); however, if we do this we may as well work only with the degree.

The rotation about a sphere $S^2$ in $R^3$ was defined by Alexander (1922) by means of an integral

$$\iint \begin{vmatrix} x & y & z \\ x_u & y_u & z_u \\ x_v & y_v & z_v \end{vmatrix} \frac{du\,dv}{r^3}$$

(where $(x, y, z)$ is the image of a point $(u, v)$ on $S^2$ and

$$r = (x^2 + y^2 + z^2)^{\frac{1}{2}});$$

a similar integral is given for higher dimensions.

Granas (1961, 1962) avoids giving a numerical value to the rotation by considering 'essential' vector fields, that is, fields with non-zero rotation. Klee (1960) extends parts of Granas's work to spaces which need not be locally convex.

## 10.2  The degree for mappings of spheres

Consider a simple closed curve $C$ in $R^2$ and a mapping $g$ of $C$ onto $S^1$. In simple cases the following facts are clear from a diagram (see for instance figure 8).

10.2.1   Most points of $S^1$ are covered a finite number of times $n_1$ in the positive sense and a finite number of times $n_2$ in the negative sense.

10.2.2   The number $n = n_1 - n_2$ is the same at all points where $n_1$ and $n_2$ exist.

10.2.3   If $g(x) = f(x)/\|f(x)\|$ for a non-zero vector field $f$ on $C$, then the rotation of $f$ about $C$ is exactly $n$.

DEFINITION 10.2.4   The number $n$ is called the *degree* of the mapping $g$.

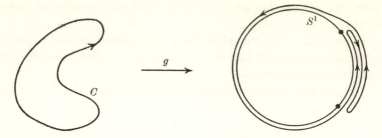

Fig. 8. A mapping of degree 1 of $C$ onto $S^1$.

We can generalise properties 10.2.1 and 10.2.2 and definition 10.2.4 to mappings of spheres in $R^n$ (indeed, to mappings of closed orientable manifolds onto orientable manifolds; see Brouwer (1910)). We can then use 10.2.3 to define the rotation about a sphere in $R^n$.

Another definition of the degree (for a mapping $g$ of $S^n$) is: $g$ induces a mapping $\tilde{g}$ of $H_n(S^n)$ into $H_n(S^n)$, that is, of $Z$ into $Z$. Thus $\tilde{g}(1)$ is some integer $n$. Define $n$ to be the degree of $g$.

We mention the main properties of the degree (for mappings of $S^n$).

10.2.5   *The degree is an integer* .

10.2.6   *The degree is unaltered by continuous deformation (homotopy) of the mapping.*

10.2.7   $\deg(ST) = \deg S \deg T$.

10.2.8   (i) $\deg I = 1$, (ii) $\deg(-I) = (-1)^{n+1}$.

Observe that 10.2.5 to 10.2.8 (i) are obvious from the homology definition. If we use the covering definition, 10.2.5, 10.2.7 and 10.2.8 (i) are obvious, 10.2.6 is intuitively plausible from figure 8, and 10.2.8 (ii) can be obtained as follows. The mapping $x \to -x$ can be obtained by $(n+1)$ reflections in the coordinate hyperplanes of $R^{n+1}$; thus orientation is reversed $(n+1)$ times.

From these properties we will derive a few fixed point theorems; many others, for spheres and projective spaces, are presented by Whittlesey (1963).

LEMMA 10.2.9   (i) *If $f$ maps $S^n$ into $S^n$ and has no fixed point, then* $\deg f = (-1)^{n+1}$. (ii) *If $-f$ has no fixed point then* $\deg f = 1$.

*Proof.* (i) The function

$$\phi(x,t) = (tf(x)+(1-t)(-x))/\|tf(x)+(1-t)(-x)\|$$

gives a homotopy between $f$ and $-I$; thus 10.2.6 and 10.2.8 give the result. (ii) Here $-f$ is homotopic to $-I$, so that $f$ is homotopic to $I$, and the result follows as before. $\square$

**Theorem 10.2.10**   *Let $f$ be a continuous mapping of $S^{2n}$ into $S^{2n}$. Then either $f$ or $-f$ has a fixed point (hence $f^2$ has a fixed point).*

*Proof.* If neither $f$ nor $-f$ has a fixed point, $\deg f$ must be both 1 and $-1$ by the lemma, which is impossible. $\square$

**Theorem 10.2.11** (*Brouwer*)   *Let $f$ be a continuous one–one mapping of $S^n$ onto $S^n$. Then $f$ has a fixed point if either*
   (i) *$n$ is even and $f$ preserves orientation, or*
   (ii) *$n$ is odd and $f$ reverses orientation.*

*Proof.* If $f$ preserves orientation, $\deg f = 1$; if $f$ reverses orientation, $\deg f = -1$. The result follows from 10.2.9 (i). $\square$

## 10.3   The degree for mappings of open sets

Consider an open subset $\mathcal{M}$ of a Banach space $\mathcal{V}$, and a continuous mapping $F$ of $\mathcal{M}$ into $\mathcal{V}$. For the moment, assume that $\mathcal{M}$ is bounded and that $\mathcal{V} = R^n$. We write $\partial\mathcal{M}$ for the boundary of $\mathcal{M}$. We illustrate a particular case in $R^2$ (figure 9).

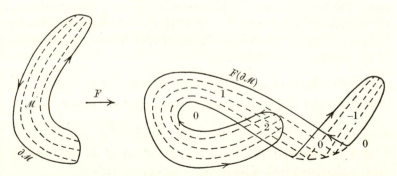

Fig. 9. A mapping of $\mathcal{M}$ by $F$. The value of the degree in each region is shown.

We observe that in this particular case:

10.3.1   The image of $\partial \mathcal{M}$ divides $\mathcal{V}$ into (connected) regions $\mathcal{U}_i$; in each region the algebraic number of times that a point is covered is constant (except at relatively few exceptional points, where the image is 'folded').

It can be shown that property 10.3.1 is true for all simplicial mappings (see for instance Cronin (1964)) or for all differentiable mappings (Nagumo, 1951*a*). In both these cases the exceptional points in a region $\mathcal{U}_i$ form a set of lower dimension, hence of measure zero.

DEFINITION 10.3.2   Let $F$ be a simplicial mapping (or a differentiable mapping) of $\mathcal{M}$ into $R^n$. The *degree* of $F$ (with respect to $\mathcal{M}$) at a point $x$ in $R^n - F(\partial \mathcal{M})$ is the algebraic number of times that (almost all) points are covered, in the region $\mathcal{U}_i$ containing $x$. This integer is written $\deg(F, \mathcal{M}, x)$, or

$$\deg(F, \mathcal{M}, \mathcal{U}_i).$$

The degree of an arbitrary continuous mapping is defined by approximating by simplicial mappings (or differentiable mappings) (see the references given above).

If $\mathcal{M}$ is an unbounded open set we must assume that $F(x) - x$ is bounded on $\mathcal{M}$; in other words, that $F - I$ is compact.

If $\mathcal{V}$ is infinite-dimensional we must approximate $F$ by finite-dimensional mappings; this requires some compactness assumption. The usual theory (until recently the only theory) defines the degree for mappings $F$ such that $F - I$ is compact. See for instance Nagumo (1951*b*) or Cronin (1964). The main properties of the degree are the following.

10.3.3   $\deg(F, \mathcal{M}, x)$ *is an integer, defined if* $x \notin F(\partial \mathcal{M})$.

10.3.4   *If* $F = I$ *then* $\deg(F, \mathcal{M}, x) = 1$ *if* $x \in \mathcal{M}$ *and*
$$\deg(F, \mathcal{M}, x) \quad = 0 \quad \text{if} \quad x \notin \overline{\mathcal{M}}.$$

10.3.5   $\deg(F, \cup \mathcal{M}_i, x) = \Sigma \deg(F, \mathcal{M}_i, x)$ *if the* $\mathcal{M}_i$ *are disjoint regions and both sides of the equation are defined.*

10.3.6   *If* $\deg(F, \mathcal{M}, x) \neq 0$ *then* $x \in F(\mathcal{M})$.

10.3.7 (*Homotopy*)   If $f(t, x)$ *is a continuous compact mapping of* $[0, 1] \times \mathcal{M}$ *into* $\mathcal{V}$, *and if* $x - f(t, x) \neq y$ *for* $x$ *in* $\partial \mathcal{M}$, *then*

$$\deg(I - f(0, \cdot), \mathcal{M}, y) = \deg(I - f(1, \cdot), \mathcal{M}, y).$$

(As in 4.3.1, we say that $f(0, \cdot)$ is $fp$-homotopic to $f(1, \cdot)$.)

10.3.8 (*Multiplication*)   If $F(\partial \mathcal{M})$ *divides* $\mathcal{V}$ *into regions* $\mathcal{U}_i$ *and if* $G(F(\partial \mathcal{M}))$ *divides* $\mathcal{V}$ *into regions* $\mathcal{W}_j$, *then*

$$\deg(GF, \mathcal{M}, \mathcal{W}_j) = \sum_i \deg(F, \mathcal{M}, \mathcal{U}_i) \deg(G, \mathcal{U}_i, \mathcal{W}_j).$$

As in § 10.2, all these properties, with the exception of homotopy, are easily seen to be true (at least in 'reasonable' cases). To make 10.3.7 plausible, consider figure 9. Let the mapping be continuously altered from $F_0 = I - f(0, \cdot)$ to $F_1 = I - f(1, \cdot)$. If $y$ lies in a region which is covered $k$ times (algebraically) by $F_0$ and if the boundary curve $F_t(\partial \mathcal{M})$ never passes over $y$, then $y$ should still be covered $k$ times by $F_1$. (Leray and Schauder suggest that the region $\mathcal{M}$ as well as the mapping $F$ can be continuously altered.)

If $F$ has no zeros in $\mathcal{M}$ except $x_1, \ldots, x_n$ then by 10.3.5 and 10.3.6,

$$\deg(F, \mathcal{M}, 0) = \Sigma \deg(F, \mathcal{M}_i, 0),$$

where the $\mathcal{M}_i$ are small neighbourhoods of the $x_i$. Leray and Schauder (1934) show how to calculate the degree using this formula and an expression for $\deg(F, \mathcal{M}_i, 0)$ in terms of the Fréchet derivative of $F$ at $x_i$. For practical purposes the degree is evaluated by using 10.3.7 and the fact that a homeomorphism has degree $\pm 1$.

*Generalisations*. Browder (1968) defines the degree for mappings of the form $H - T$, where $H$ is a homeomorphism and $T$ is compact (also, for more general combinations of $H$ and $T$). Properties 10.3.3 to 10.3.7 are obtained, but 10.3.8 is false in this case. The degree can also be defined for other classes of mappings. For further discussion, applications and references see Browder (1973).

The degree $\deg(f, \mathcal{M}, x)$ can be defined in cases where $\mathcal{M}$ is a subset of a space (for example, a manifold) which is not a vector space; see Browder (1960$b$).

*Applications* of degree theory to fixed point theorems usually involve one of the following theorems.

**THEOREM 10.3.9**   *If* $\deg(I - T, \mathcal{M}, 0) \neq 0$ *then* $T$ *has a fixed point in* $\mathcal{M}$.

*Proof.* By 10.3.6, we have $0 \in (I - T)\mathcal{M}$; thus $0 = y - Ty$ for some $y$ in $\mathcal{M}$. □

**THEOREM 10.3.10** (*Leray–Schauder*, 1934)   *If*

$$\deg(I - T_0, \mathcal{M}, 0) \neq 0$$

*and* $T_1$ *is fp-homotopic to* $T_0$ (*that is, homotopic under a compact homotopy with no fixed points on* $\partial \mathcal{M}$), *then* $T_1$ *has a fixed point in* $\mathcal{M}$.

*Proof.* By 10.3.7,

$$\deg(I - T_1, \mathcal{M}, 0) = \deg(I - T_0, \mathcal{M}, 0) \neq 0,$$

so that 10.3.9 gives a fixed point for $T_1$. □

We will now show how some of our earlier fixed point theorems can be derived from 10.3.9 and 10.3.10.

**A PROOF OF BROUWER'S THEOREM 2.1.11**   Write $\mathcal{M}$ for the unit ball in $R^n$. Suppose that $U$ maps $\mathcal{M}$ into $\mathcal{M}$.

*Case 1.* $U$ has a fixed point on $\partial \mathcal{M}$.

*Case 2.* $U$ has no fixed point on $\partial \mathcal{M}$. Then $f(t, x) = tUx$ gives a *fp*-homotopy between $U$ and 0. Thus

$$\deg(I - U, \mathcal{M}, 0) = \deg(I, \mathcal{M}, 0) = 1,$$

by 10.3.4. By 10.3.9, $U$ has a fixed point in $\mathcal{M}$.

Thus in either case, $U$ has a fixed point in $\mathcal{M}$. □

This argument (slightly adapted) also proves Rothe's theorem and the fact that
$$\deg(I - U, \mathcal{M}, 0) = 1,$$

if $U$ satisfies the conditions of Rothe's theorem and has no fixed points on $\partial \mathcal{M}$. In the Browder–Potter theorem 4.3.3 the initial mapping $U_0$ satisfied the conditions of Rothe's theorem; thus 4.3.3 is a special case of 10.3.10.

## 10.4 The index and Lefschetz number of a mapping

Let $T$ be a compact mapping of a bounded open subset $\mathcal{M}$ (of a locally convex space $\mathcal{V}$) into $\mathcal{V}$.

**DEFINITION 10.4.1** The *index* of $T$, written $i(T)$, is $\deg(I - T, \mathcal{M}, 0)$.

In theorems 10.3.9 and 10.3.10 we could have said that $i(T) \neq 0$ instead of that $\deg(I - T, \mathcal{M}, 0) \neq 0$. However, we will continue to speak of the degree rather than the index.

10.4.2 The *Lefschetz number* of a self-mapping $f$ of a compact set $\mathcal{N}$ (usually a polyhedron or an absolute neighbourhood retract) is an integer $L(f)$ whose most important property is that

$$L(f) \neq 0 \Rightarrow f \text{ has a fixed point in } \mathcal{N}.$$

We can define $L(f)$ in several ways. Leray (1950) suggests:

**DEFINITION 10.4.3** Embed $\mathcal{N}$ in a locally convex space. Let $\mathcal{M}$ be an open neighbourhood of $\mathcal{N}$ and let $r$ be a retraction of $\mathcal{M}$ onto $\mathcal{N}$. Then we put

$$L(f) = \deg(I - fr, \mathcal{M}, 0).$$

We will not show that $L(f)$ is well defined by 10.4.3, or discuss its properties. However, 10.4.2 obviously follows from our discussion of the degree.

10.4.4 Lefschetz's own definition of the number $L(f)$ has the form

$$L(f) = \Sigma(-1)^n \operatorname{Trace} T_n,$$

where the $T_n$ are endomorphisms of the homology groups of $\mathcal{N}$, induced by $f$. For a discussion of $L(f)$ starting from this definition, see Spanier (1966) or Brown (1971). (Spanier deals only with the case of polyhedra.) With this definition of $L(f)$, 10.4.2 is known as the *Lefschetz fixed point theorem*.

In certain cases it can be shown that $L(f) \neq 0$ for all continuous mappings of $\mathcal{N}$ into $\mathcal{N}$. If so, $\mathcal{N}$ has the fixed point property. This approach yields the fixed point property for the $n$-ball or the projective spaces of even dimension. If $L(f) = 0$, when calculated by 10.4.4 using homology groups with integral coefficients, a non-zero value may sometimes be obtained by using a different coefficient group. For more details, see Fadell (1970).

## Exercises

1. (*Alexander*, 1922)   Let $\mathcal{M} = \mathcal{D} - \bigcup_1^n \mathcal{D}_i^o$ where the $\mathcal{D}_i$ are non-overlapping closed discs in the interior of a closed disc $\mathcal{D}$. Let $T$ be a continuous map of $\mathcal{M}$ into $\mathcal{M}$ which maps each boundary circle $\partial\mathcal{D}$ or $\partial\mathcal{D}_i$ into itself, preserving the orientation of each of these circles. Then if $n \geqslant 2$, $T$ has a fixed point in $\mathcal{M}$. (Consider rot $(T-I, \mathcal{D})$ and rot $(T-I, \mathcal{D}_i)$.)

2. Show that for the mapping $F : z \to z^n$ of the unit disc $\mathcal{M}$ we have $\deg(F, \mathcal{M}, 0) = \text{rot}(F, \partial\mathcal{M})$ and that the same relation holds in figure 9, wherever the origin may lie, provided that $0 \notin F(\partial\mathcal{M})$.

# 11. *Further topics*

## More or less general surveys

Some useful surveys are mentioned in the introduction. We should also mention Nemyckii (1936), Leray (1950), Newman (1952), Krasnoselskii (1954), Granas (1961, 1962), Bonsall (1962), Cronin (1964) and Edwards (1965).

## Measure-preserving mappings

Some theorems give fixed points for measure-preserving mappings, where continuity alone would not be enough. For homeomorphisms (preserving orientation) of the open unit disc, see Bourgin (1968). For motions of the annulus between two circles, see Poincaré (1912) and Birkhoff (1913, 1927). (The motion must advance points on the one circle and regress points on the other circle.) Both Poincaré and Birkhoff give applications to periodic motions of dynamical systems.

## Partially ordered spaces

For fixed point theorems under various conditions related to the ordering in these spaces, and for applications see for instance Kantorovitch (1939), Vulikh (1967), Krasnoselskii (1964) or Bonsall (1962).

## Computation of fixed points

Banach's theorem 1.2.2 provides an algorithm for the calculation of fixed points of contraction mappings. Various authors have given algorithms applicable to non-expansive mappings. See for instance Browder and Petryshyn (1966), Halpern (1967),

Reinermann (1971*b*) or Kaneil (1971). For continuous mappings of a simplex, simplicial subdivision clearly gives a method of finding $\epsilon$-fixed points; Scarf (1967) gives an algorithm. If we also consider the rotation about a subsimplex we can (in principle) approximate actual fixed points; see Vertgeim (1970). A simple method is to make successive bisections, at each stage taking the subsimplex about which the rotation is nonzero. However, $\epsilon$-fixed points of $T$ on the boundary of a subsimplex might prevent computation of the rotation of $I - T$ about that subsimplex. If this occurred we would find only the $\epsilon$-fixed points.

# Bibliography

Alexander, J. W. (1922). On transformations with invariant points. *Trans. Amer. Math. Soc.* **23**, 89–95.

Alexandroff, P. and Hopf, H. (1935). *Topologie*, Vol. i. Springer, Berlin.

Bailey, D. F. (1966). Some theorems on contractive mappings. *J. Lond. Math. Soc.* **41**, 101–6.

Banach, S. (1922). Sur les opérations dans les ensembles abstraits et leur application aux équations intégrales. *Fund. Math.* **3**, 133–81.

—— (1932). *Théorie des opérations linéaires*. Warsaw.

Bass, R. (1958). On nonlinear repulsive forces. In Lefschetz, S. (ed.), *Contributions to the theory of nonlinear oscillations*, Vol. iv, pp. 201–11. Princeton University Press.

Begle, E. G. (1950). A fixed point theorem. *Annals of Math.* (2) **51**, 544–50.

Belluce, L. P. and Kirk, W. A. (1966). Fixed point theorems for families of contraction mappings. *Pacific J. Math.* **18**, 213–17.

Bielecki, A. (1956). Une remarque sur la méthode de Banach–Caccioppoli–Tikhonov. *Bull. Acad. Polon. Sci.* **4**, 261–8.

Bing, R. H. (1969). The elusive fixed point property. *Amer. Math. Monthly* **76**, 119–31.

Birkhoff, G. D. (1913). Proof of Poincaré's geometric theorem. *Trans. Amer. Math. Soc.* **14**, 14–22.

—— (1927). *Dynamical systems*. Amer. Math. Soc., Providence, R.I.

Birkhoff, G. D. and Kellogg, O. D. (1922). Invariant points in function space. *Trans. Amer. Math. Soc.* **23**, 96–115.

Bohl, P. (1904). Über die Bewegung eines mechanischen Systems in der Nähe einer Gleichgewichtslage. *J. Reine Angew. Math.* **127**, 179–276.

Bohnenblust, H. F. and Karlin, S. (1950). On a theorem of Ville. In Kuhn, H. W. and Tucker, A. W. (eds.), *Contributions to the theory of games*, Vol. i, pp. 155–60. Princeton University Press.

Bonsall, F. F. (1962). *Lectures on some fixed point theorems of functional analysis*. Tata Institute, Bombay.

Bourbaki, N. (1955). *Espaces vectoriels topologiques*. Hermann, Paris.

Bourgin, D. G. (1968). Homeomorphisms of the open disc. *Studia Math.* **31**, 433–8.

Boyce, W. M. (1969). Commuting functions with no common fixed point. *Trans. Amer. Math. Soc.* **137**, 77–92.

Brouwer, L. E. J. (1910). Über Abbildung von Mannigfaltigkeiten. *Math. Ann.* **71**, 97–115.

—— (1952). An intuitionist correction of the fixed-point theorem on the sphere. *Proc. Roy. Soc. London* (A) **213**, 1–2.

Browder, F. E. (1959). On a generalization of the Schauder fixed point theorem. *Duke Math. J.* **26**, 291–303.

—— (1960a). On continuity of fixed points under deformation of continuous mappings. *Summa Brasil. Math.* **4**, 183–90.

[ 87 ]

Browder, F. E. (1960b). On the fixed point index for continuous mappings of locally connected spaces. *Summa Brasil. Math.* **4**, 253–93.

(1965a). Fixed point theorems on infinite dimensional manifolds. *Trans. Amer. Math. Soc.* **119**, 179–94.

(1965b). Fixed point theorems for noncompact mappings in Hilbert space. *Proc. Nat. Acad. Sci. U.S.A.* **53**, 1272–6.

(1965c). Existence of periodic solutions for nonlinear equations of evolution. *Proc. Nat. Acad. Sci. U.S.A.* **53**, 1100–3.

(1965d). Nonexpansive nonlinear operators in a Banach space. *Proc. Nat. Acad. Sci. U.S.A.* **54**, 1041–4.

(1966). *Problèmes non-linéaires*. Lecture notes. University of Montreal.

(1968). Topology and nonlinear functional equations. *Studia Math.* **31**, 189–204.

(1973). *Nonlinear operators and nonlinear equations of evolution in Banach spaces*, being *Nonlinear functional analysis*, Vol. ii. Amer. Math. Soc., New York.

Browder, F. E. and Petryshyn, W. V. (1966). The solution by iteration of nonlinear functional equations in Banach spaces. *Bull. Amer. Math. Soc.* **72**, 571–5.

Brown, R. F. (1971). *The Lefschetz fixed point theorem*. Glenview, Illinois.

Bruck, R.E. (1973). Properties of fixed-point sets of nonexpansive mappings in Banach Spaces. *Trans. Amer. Math. Soc.* **179**, 251–62.

Caccioppoli, R. (1930). Un teorema generale sull'esistenza di elementi uniti in una trasformazione funzionale. *Rend. Accad. Lincei* (6) **11**, 794–9.

Copson, E. T. (1968). *Metric spaces*. Cambridge University Press.

Courant, R. and Hilbert, D. (1962). *Methods of mathematical physics*, Vol. ii. Interscience, New York.

Courant, R. and Robbins, H. (1961). *What is mathematics?* Oxford University Press.

Cronin, J. (1964). *Fixed points and topological degree in nonlinear analysis*. Amer. Math. Soc., New York.

Day, M. M. (1958). *Normed linear spaces*. Springer, Berlin.

(1961). Fixed point theorems for compact convex sets. *Ill. J. Math.* **5**, 585–90. (Corrected in (1964), *Ill. J. Math.* **8**, 713.)

DeMarr, R. (1963a). Common fixed points for commuting contraction mappings. *Pacific J. Math.* **13**, 1139–41.

(1963b). A common fixed point theorem for commuting mappings. *Amer. Math. Monthly* **70**, 535–7.

Dugundji, J. (1951). An extension of Tietze's theorem. *Pacific J. Math.* **1**, 353–67.

(1958). Absolute neighbourhood retracts and local connectedness in arbitrary metric spaces. *Compos. Math.* **13**, 229–46.

Dunford, N. and Schwartz, J. T. (1958). *Linear operators*, Vol. i. Interscience, New York.

(1963). *Linear operators*, Vol. ii. Interscience, New York.

Edelstein, M. (1961). An extension of Banach's contraction principle. *Proc. Amer. Math. Soc.* **12**, 7–10.

(1962). On fixed points and periodic points under contraction mappings. *J. London Math. Soc.* **37**, 74–9.

(1969). A short proof of a theorem of Janos. *Proc. Amer. Math. Soc.* **20**, 509–10.

Edwards, R. E. (1965). *Functional analysis : theory and applications* (Chapter 3). Holt, Rinehart and Winston, New York.

Eilenberg, S. and Montgomery, D. (1946). Fixed point theorems for multivalued transformations. *Amer. J. Math.* **68**, 214–22.

Fadell, E. (1970). Recent results in the fixed point theory of continuous maps. *Bull. Amer. Math. Soc.* **76**, 10–29.

Fleischman, W. M. (ed.) (1970). *Set-valued mappings, selections and topological properties of $2^X$.* Lecture notes. Springer, Berlin.

Folkman, J. H. (1966). On functions that commute with full functions. *Proc. Amer. Math. Soc.* **17**, 383–6.

Fort, M. K. (1954). Open topological discs in the plane. *J. Indian Math. Soc.* **18**, 23–6.

Glicksberg, I. (1952). A further generalisation of the Kakutani fixed-point theorem, with application to Nash equilibrium points. *Proc. Amer. Math. Soc.* **3**, 170–4.

Goebel, K. (1969). An elementary proof of the fixed-point theorem of Browder and Kirk. *Michigan Math. J.* **16**, 381–3.

Górniewicz, L. and Granas, A. (1970). Fixed point theorems for multi-valued mappings of the absolute neighbourhood retracts. *J. Math. Pures et Appl.* **49**, 381–95.

Granas, A. (1959). Theorem on antipodes and theorems on fixed points for a certain class of multi-valued mappings in Banach spaces. *Bull. Acad. Pol. Sci.* **7**, 271–5.

(1961). *Introduction to topology of functional spaces.* Lecture notes, University of Chicago.

(1962). *The theory of compact vector fields and some of its applications to topology of functional spaces* (i). Warsaw.

Graves, L. M. (1946). *The theory of functions of real variables.* McGraw-Hill, New York.

Greenleaf, F. P. (1969). *Invariant means on topological groups.* Van Nostrand, New York.

Güssefeldt, G. (1970). Zwei Methoden zum Nachweis von periodischen Lösungen bei gewöhnlichen Differentialgleichungen. *Math. Nachr.* **48**, 141–51.

Halpern, B. (1967). Fixed points of nonexpanding maps. *Bull. Amer. Math. Soc.* **73**, 957–61.

Harris, W. A., Sibuya, Y. and Weinberg, L. (1969). Holomorphic solutions of linear differential systems. *Arch. Rat. Mech. Anal.* **35**, 245–8.

Hausdorff, F. (1914). *Grundzüge der Mengenlehre.* Leipzig.

Hirsch, M. W. (1963). A proof of the nonretractability of a cell onto its boundary. *Proc. Amer. Math. Soc.* **14**, 364–5.

Huneke, J. P. (1969). On common fixed points of commuting continuous functions on an interval. *Trans. Amer. Math. Soc.* **139**, 371–81.

Janos, L. (1967). A converse of Banach's contraction theorem. *Proc. Amer. Math. Soc.* **18**, 287–9.

Jones, G. S. (1965). Stability and asymptotic fixed point theory. *Proc. Nat. Acad. Sci.* **53**, 1262–4.

Kakutani, S. (1938). Two fixed-point theorems concerning bicompact convex sets. *Proc. Imp. Acad. Tokyo* **14**, 242–5.

(1941). A generalisation of Brouwer's fixed point theorem. *Duke Math. J.* **8**, 457–9.

(1943). Topological properties of the unit sphere of a Hilbert space. *Proc. Imp. Acad. Tokyo* **19**, 269–71.

Kaniel, S. (1971). Construction of a fixed point for contractions in Banach space. *Israel J. Math.* **9**, 535–40.

Kantorovitch, L. (1939). The method of successive approximations for functional equations. *Acta Math.* **71**, 63–97.

Karlin, S. S. (1959). *Mathematical methods and theory in games, programming and economics*. Vols. I, II. Addison-Wesley, London.

Kinoshita, S. (1953). On some contractible continua without the fixed point property. *Fund. Math.* **40**, 96–8.

Kirk, W. A. (1965). A fixed point theorem for mappings which do not increase distance. *Amer. Math. Monthly* **72**, 1004–6.

(1970). Fixed point theorems for nonexpansive mappings. In Browder, F. E. (ed.), *Nonlinear functional analysis*, pp. 162–8. Amer. Math. Soc., New York.

Klee, V. L. (1955). Some topological properties of convex sets. *Trans. Amer. Math. Soc.* **78**, 30–45.

(1960). Leray–Schauder theory without local convexity. *Math. Ann.* **141**, 286–96.

Knaster, B., Kuratowski, C. and Mazurkiewicz, S. (1929). Ein Beweis des Fixpunktsatzes für *n*-dimensionale Simplexe. *Fund. Math.* **14**, 132–7.

Kolmogorov, A. N. and Fomin, S. J. (1957). *Functional analysis*. Vol. I. Graylock, Rochester.

Krasnoselskii, M. A. (1954). Some problems of nonlinear analysis. (Russian) *Uspehi Mat. Nauk* (N.S.) **9**, No. 3 (61), 57–114 = *Amer. Math. Soc. Transl.* (2) **10** (1958), 345–409.

(1964). *Positive solutions of operator equations*. Groningen.

Kuratowski, C. (1933). *Topologie*, Vol. I. Warsaw.

(1968). *Topology*, Vol. II. Hafner, New York.

Ky Fan (1952). Fixed point and minimax theorems in locally convex topological linear spaces. *Proc. Nat. Acad. Sci. U.S.* **38**, 121–6.

(1961). A generalization of Tychonoff's fixed point theorem. *Math. Ann.* **142**, 305–10.

Lasota, A. and Opial, Z. (1965). An application of the Kakutani–Ky Fan theorem in the theory of ordinary differential equations. *Bull. Acad. Polon. Sci.* **13**, 781–6.

Lefschetz, S. (1930). *Topology*. Amer. Math. Soc., New York.

(1942). *Algebraic topology*. Amer. Math. Soc., New York.

Leray, J. (1950). La théorie des points fixes et ses applications en analyse. *Proc. Int. Congress of Math.*, Vol. II, pp. 202–8. Amer. Math. Soc., New York.

Leray, J. and Schauder, J. (1934). Topologie et équations fonctionnelles. *Ann. Ecole Norm.* (3) **51**, 45–78.

Lipschitz, R. (1876). Sur la possibilité d'intégrer complètement un système donné d'équations différentielles. *Bull. Sci. Math.* **10**, 149–59.

Markov, A. (1936). Quelques théorèmes sur les ensembles abéliens. *C. R. Acad. Sci. URSS* (N.S.) **1**, 311–13.

Maunder, C. R. F. (1970). *Algebraic topology*. Von Nostrand, London.

Meyers, P. R. (1967). A converse to Banach's contraction theorem. *J. Res. Nat. Bur. Stand.* **71**B, 73–6.

(1970). Contractifiable semigroups. *J. Res. Nat. Bur. Stand.* **74**B, 315–22.

Miranda, C. (1955). *Equazione alle derivate parziali di tipo ellittico*. Springer, Berlin.

Nagumo, M. (1951*a*). A theory of degree of mapping based on infinitesimal calculus. *Amer. J. Math.* **73**, 485–96.

(1951*b*). Degree of mapping in convex linear topological spaces. *Amer. J. Math.* **73**, 497–511.

Namioka, I. and Asplund, E. (1967). A geometric proof of Ryll-Nardzewski's fixed point theorem. *Bull. Amer. Math. Soc.* **73**, 443–5.

Nemyckii, V. V. (1936). The fixed point method in analysis. (Russian.) *Uspehi Mat. Nauk.* (N.S.) **1**, 141–74 = *Amer. Math. Soc. Transl.* (2) **34** (1963), 1–37.

von Neumann, J. (1929). Zur allgemeinen Theorie des Masses. *Fund. Math.* **13**, 73–116.

— (1935). Über ein ökonomisches Gleichungssystem und eine Verallgemeinerung des Brouwerschen Fixpunktsatzes. *Ergebnisse eines Mathematischen Kolloquiums* **8**, 73–83.

Newman, M. H. A. (1952). Fixed point and coincidence theorems. *J. Lond. Math. Soc.* **27**, 135–40.

Nirenberg, L. (1953). On nonlinear elliptic partial differential equations and Hölder continuity. *Comm. Pure Appl. Math.* **6**, 103–56. (Addendum, p. 395.)

Poincaré, H. (1912). Sur un théorème de Géométrie. *Rend. Circ. Mat. Palermo* **33**, 375–407.

Potter, A. J. B. (1973). An elementary version of the Leray–Schauder theorem. *J. Lond. Math. Soc.* (to appear).

Reinermann, J. (1971 *a*). Fixpunktsätze vom Krasnoselski–Typ. *Math. Zeit.* **119**, 339–44.

— (1971 *b*). Approximation von Fixpunkten. *Studia Math.* **39**, 1–15.

Rothe, E. (1937). Zur Theorie der topologischen Ordnung und der Vektorfelder in Banachschen Räumen. *Compos. Math.* **5**, 177–96.

Ryll-Nardzewski, C. (1966). On fixed points of semi-groups of endomorphisms of linear spaces. *Proceedings of the Fifth Berkeley Symposium on Statistics and Probability*, Vol. ɪɪ. Berkeley.

Sadovskii, B. N. (1967). A fixed point principle. *Funktsional Anal. i Priložen* **1**, 74–6 = *Func. Anal. and Applications* **1**, 151–3.

— (1972). Limit-compact and condensing operators. *Russ. Math. Surveys* **27**, 85–155 = *Uspekhi Mat. Nauk.* **27**, 81–146.

Scarf, H. (1967). The approximation of fixed points of a continuous mapping. *SIAM J. Appl. Math.* **15**, 1328–43.

Schaefer, H. (1955). Über die Methode der a priori Schranken. *Math. Ann.* **129**, 415–16.

Schauder, J. (1927). Zur Theorie stetiger Abbildungen in Funktionalräumen. *Math. Z.* **26**, 47–65 and 417–31.

— (1930). Der Fixpunktsatz in Funktionalräumen. *Studia Math.* **2**, 171–80.

— (1932). Über den Zusammenhang zwischen der Eindeutigkeit und Lösbarkeit partieller Differentialgleichungen zweiter Ordnung von Elliptischen Typus. *Math. Ann.* **106**, 661–721.

Smart, D. R. (1967). Fixed points in a class of sets. *Pacific J. Math.* **23**, 163–5.

— (1973 *a*). Fixed points of homotopic mappings. To appear.

— (1973 *b*). Fixed point theorems of many-valued mappings. To appear.

Smithson, R. E. (1965). Changes of topology and fixed points for multi-valued functions. *Proc. Amer. Math. Soc.* **16**, 448–54.

— (1972). Multifunctions. *Nieuw Arch. Wisk.* (3) **20**, 31–53.

Spanier, E. (1966). *Algebraic topology.* McGraw-Hill, New York.

Stokes, A. (1960). The applications of a fixed point theorem to a variety of non-linear stability problems. In Lefschetz, S. (ed.), *Contributions to the theory of nonlinear oscillations*, Vol. v. Princeton University Press.

Tychonoff, A. (1935). Ein Fixpunktsatz. *Math. Ann.* **111**, 767–76.

van der Walt, T. (1963). *Fixed and almost fixed points.* Mathematical Centrum, Amsterdam.

Vertgeim, B. A. (1970). On an approximate determination of the fixed points of continuous mappings. *Dokl. Akad. Nauk. SSSR* **191** = *Soviet Math. Dokl.* **11**, 295–8.

Vidossich, G. (1971). Applications of topology to analysis. *Confer. Sem. Mat. Univ. Bari* **126**, 1–62.

Vulikh, B. Z. (1967). *Introduction to the theory of partially ordered spaces.* Groningen.

Whittlesey, P. (1963). Fixed points and antipodal points. *Amer. Math. Monthly* **70**, 807–21.

# Index